Artificial Intelligence in Highway Location and Alignment Optimization

Applications of Genetic Algorithms in Searching, Evaluating, and Optimizing Highway Location and Alignments

Other Related Titles from World Scientific

What is Artificial Intelligence?
A Conversation between an AI Engineer and a Humanities Researcher
by Suman Gupta and Peter H. Tu
ISBN: 978-1-78634-863-0

Artificial Intelligence in Highway Location and Alignment Optimization

Applications of Genetic Algorithms in Searching, Evaluating, and Optimizing Highway Location and Alignments

Min-Wook Kang

University of South Alabama, USA

Paul Schonfeld

University of Maryland, College Park, USA

World Scientific

NEW JERSEY · LONDON · SINGAPORE · BEIJING · SHANGHAI · HONG KONG · TAIPEI · CHENNAI · TOKYO

Published by

World Scientific Publishing Co. Pte. Ltd.
5 Toh Tuck Link, Singapore 596224
USA office: 27 Warren Street, Suite 401-402, Hackensack, NJ 07601
UK office: 57 Shelton Street, Covent Garden, London WC2H 9HE

Library of Congress Cataloging-in-Publication Data
Names: Kang, Min-Wook, author. | Schonfeld, Paul, author.
Title: Artificial intelligence in highway location and alignment optimization : applications of
    genetic algorithms in searching, evaluating, and optimizing highway location and alignments /
    Min-Wook Kang, University of South Alabama, USA,
    Paul Schonfeld, University of Maryland, College Park, USA.
Description: Hackensack, NJ : World Scientific, [2020] |
    Includes bibliographical references and index.
Identifiers: LCCN 2019057609 | ISBN 9789813272804 (hardcover) | ISBN 9789813272811 (ebook)
Subjects: LCSH: Roads--Design and construction--Mathematical models. |
    Highway planning--Mathematical models. | Genetic algorithms. |
    Artificial intelligence--Engineering applications. | Mathematical optimization.
Classification: LCC TE175 .K25 2020 | DDC 625.7/25028563--dc23
LC record available at https://lccn.loc.gov/2019057609

British Library Cataloguing-in-Publication Data
A catalogue record for this book is available from the British Library.

For any available supplementary material, please visit
https://www.worldscientific.com/worldscibooks/10.1142/11059#t=suppl

About the Authors

Min-Wook Kang is an Associate Professor at the University of South Alabama. Prior to joining it, he received professional preparation from transportation research centers, universities, and consulting companies as a research assistant and transportation engineer. He holds a Ph.D. in Civil and Environmental Engineering (Transportation Engineering focus) from the University of Maryland — College Park, and is a registered professional engineer in the state of Maryland. His areas of interest include traffic operation, safety, and congestion management of highways and intersections, Artificial Intelligence (AI)-based optimization for transport system design and other safety-related research in transportation systems. He has over 60 publications, including 22 peer-reviewed journals. He is a member of ASCE Transportation and Development Institute and a recipient the ASCE ExCEEd Fellow Award. He established and is the founding director of Transportation Safety, Simulation, & Optimization (TSSO) Lab at the University of South Alabama.

Paul Schonfeld is a Professor of Civil Engineering at the University of Maryland, where he served for 19 years as Director of its Transportation Engineering Program. He has B.S. and M.S. degrees from the Massachusetts Institute of Technology and a Ph.D. from the University of California at Berkeley. He has experience in analyzing various transportation systems, including road networks and traffic management systems, public transportation systems, freight logistics, inland waterways, and airports. He has over 540 publications, including 167 accepted for peer-reviewed journals. He has served as Editor of the *Journal of Advanced Transportation* and of ASCE's *Journal of Transportation Engineering*. He is a Fellow of ASCE and Institute of Transportation Engineers (ITE). 26 of his Ph.D. students have accepted university faculty appointments. He is the recipient of ASCE's 2018 James Laurie Prize for career achievements in transportation engineering.

Contents

About the Authors   v

**Part I: Overview of Highway Location and Alignment
           Optimization Problem**   **1**

Chapter 1: Introduction   **3**

1.1   Background   3
1.2   Highway Alignment Optimization Problems   4
1.3   Organization of the Monograph   8

Chapter 2: Highway Cost and Constraints   **11**

2.1   Initial Construction Costs   12
2.2   Highway Maintenance Costs   13
2.3   User Costs   13
2.4   Environmental and Socio-economic Impacts   14
2.5   Highway Constraints   15
       2.5.1   Geometric Constraints   15
       2.5.2   Geographical and Environmental Constraints . .   16

**Chapter 3:   Review of Artificial Intelligence-based
               Models for Optimizing Highway Location
               and Alignment Design**   **17**

3.1   Genetic Algorithms-based Optimization Models   17
       3.1.1   Highway Alignment Optimization (HAO) Models
                by the UMD Research Team   18

3.1.2   Other Models that Use Genetic Algorithms
        for Optimizing Highway Alignments   23
3.2   Other Artificial Intelligence-based Optimization Models .   24
      3.2.1   Swarm Intelligence-based Models   24
      3.2.2   Neighborhood Search Heuristic   26
3.3   Other Methods Tried with AIs for Optimizing Highway
      Alignments   27
      3.3.1   Distance Transform   27
      3.3.2   Multi-objective Optimization   28
      3.3.3   Multiple Path Optimization   28
      3.3.4   Other Objectives Considered   28
      3.3.5   Summary   29

**Part II: Highway Alignment Optimization with Genetic
            Algorithms   31**

Chapter 4:   Modeling Highway Alignments with GAs   33

4.1   Representation of Highway Alignments with GAs   33
4.2   Modeling Horizontal Alignments   36
      4.2.1   Horizontal Alignment Generation Procedure . . .   39
4.3   Modeling Vertical Alignments   40
4.4   Modeling Highway Endpoints   42
      4.4.1   Determination of Highway Endpoints   42
      4.4.2   Highway Endpoint Determination Procedure . .   45
      4.4.3   GA Operators for Endpoint Generation   47
4.5   Modeling Highway Structures   49
      4.5.1   Small Highway Bridges for Grade Separation . .   50
      4.5.2   Structures for Highway Junction Points
              with Existing Roads   52

Chapter 5:   Highway Alignment Optimization Formulation   55

5.1   Objective Function   55
5.2   Constraints   57
5.3   Integrating GAs and Geographic Information System . .   59

5.4   Highway Cost Formulation   62
       5.4.1   Highway Agency Cost   62
       5.4.2   User Cost   68
       5.4.3   Penalty and Environmental Costs   70
       5.4.4   Life-cycle Cost   74

**Chapter 6:   Constraint Handling for Evolutionary
                Algorithms   75**

6.1   Direct Constraint Handling   78
       6.1.1   Elimination Method   78
       6.1.2   Repairing Method   78
       6.1.3   Preserving Method   79
       6.1.4   Decoding Method   79
       6.1.5   Locating the Boundary of the Feasible Regions .   80
6.2   Indirect Constraint Handling (Penalty Approaches) . . .   80
       6.2.1   Death Penalty   81
       6.2.2   Static Penalty   82
       6.2.3   Dynamic Penalty   83
       6.2.4   Adaptive Penalty   84
6.3   Handling Infeasible Solutions of GA-based Highway
       Alignment Optimization   84

**Chapter 7:   Highway Alignment Optimization Through
                Feasible Gates   87**

7.1   Research Motivation of Feasible Gates   87
7.2   Feasible Gates for Horizontal Alignments   90
       7.2.1   User-defined Horizontal Feasible Bounds   90
       7.2.2   Representation of Horizontal Feasible Gates . . . 92
       7.2.3   Horizontal Feasible Gate Determination
                Procedure   96
       7.2.4   User-defined Constraints for Guiding Feasible
                Alignments   96
7.3   Feasible Gates for Vertical Alignments   99
       7.3.1   Road Elevation Determination Procedure   100
7.4   Example Study   102
7.5   Summary   106

**Chapter 8:   Prescreening and Repairing in Highway
                Alignment Optimization   109**

8.1   Research Motivation of Prescreening and Repairing . . .   109
8.2   Prescreening and Repairing for Alignments Violating
       Design Constraints   112
       8.2.1   P&R Basic Concept   112
       8.2.2   Determination of Design Constraint Violations .   113
8.3   Example Study   117
8.4   Summary   121

**Part III: Optimizing Simple Highway Networks:
              An Extension of Genetic Algorithms-based
              Highway Alignment Optimization   123**

Chapter 9:   Overview of Discrete Network Design Problems   125

9.1   Bi-level Discrete Network Design Problems   125
       9.1.1   Upper-level DNDP   127
       9.1.2   Lower-level DNDP   127
9.2   Comparison of Highway Alignment Optimization
       and Discrete Network Design Problems   127

**Chapter 10:   Bi-level Highway Alignment Optimization
               within a Small Highway Network   131**

10.1   Bi-level HAO Concept   132
10.2   Upper Level of Bi-level HAO   134
       10.2.1   Highway Agency Cost for Bi-level HAO   135
       10.2.2   Highway User Cost for Bi-level HAO   135
       10.2.3   Penalty and Environmental Costs   144
10.3   Lower Level of Bi-level HAO   145
       10.3.1   User and System Optimal Traffic Assignment
                Problems   145
       10.3.2   Determination of Traffic Reassignment   146
10.4   Bi-level HAO Model Structure   148
10.5   Inputs Required for Lower Level of Bi-level HAO   150
       10.5.1   Highway Network and O/D Trip Matrix   150
       10.5.2   Travel Time Functions   156
10.6   Summary and Future Work   160

Chapter 11: Bi-level HAO Model Application Example   **163**

11.1   Example Description 163
11.2   Optimized Alternatives 166
11.3   Summary 169

Part IV: Highway Alignment Optimization Model
            Applications and Extensions   **171**

Chapter 12: HAO Model Application in Maryland
                Brookeville Bypass Project   **173**

12.1   Project Description 173
12.2   Data and Application Procedure 174
      12.2.1   Horizontal Map Digitization 174
      12.2.2   Vertical Map Digitization 176
      12.2.3   Tradeoffs in Map Representation
              for Environmental Issues 176
      12.2.4   Description of Model Inputs and Outputs 178
12.3   Optimization Results 180
      12.3.1   Optimized Alignments with Different Numbers
              of PI's 180
      12.3.2   Goodness Test 185
12.4   Sensitivity of Optimized Alignments to Other Major
      Input Parameters 188
      12.4.1   Sensitivity to the Model's Objective Function . . 188
      12.4.2   Sensitivity to Design Speed 191
      12.4.3   Sensitivity to Elevation Resolution 192
      12.4.4   Sensitivity to Cross-section Spacing 193

Chapter 13: HAO Model Application in US 220 Project
                in Maryland   **195**

13.1   Project Description 195
13.2   Projection Preparation 197
      13.2.1   Spatial Data 197
      13.2.2   Project Segmentation 202
      13.2.3   Geometric Design Specification 203
13.3   Optimization Results 205

Chapter 14: HAO Model Application to Maryland ICC
Project   **211**

14.1   Project Description   211
        14.1.1 Overview of the ICC Study   211
        14.1.2 Description of HAO Model Application
                to ICC Project   213
14.2   Input Data Preparation   214
        14.2.1 Road Network   214
        14.2.2 Traffic Information   214
        14.2.3 GIS Map Preparation   215
        14.2.4 Important Input Parameters   218
14.3   Optimization Results   220
        14.3.1 Determination of Traffic Reassignments   220
        14.3.2 Optimized Alignments   225
        14.3.3 Goodness Test   231

Chapter 15: Related Developments and Extensions   233

15.1   Related Developments   233
15.2   Extensions   235

Appendix A: Notation Used in the Monograph   241

Appendix B: Traffic Inputs to the HAO Model for the ICC
                Case Study   251

References   255

Index   271

Part I

Overview of Highway Location and Alignment Optimization Problem

Chapter 1

Introduction

1.1 Background

Increasing highway traffic and safety concerns often justify the construction of new highways and bypass routes or the realignment and expansion
of existing highways. Highway agencies are then challenged to locate and
design the best possible alternatives. However, unlike the rapid developments in the automobile and construction industries, search and design
processes for highway location are still carried out in a traditional process
which is more than 50 years old. Finding preferred highway alternatives
with existing methods requires considerable resources (e.g., manpower and
time). Furthermore, the agencies often face complex decisions in align-
ing a road and estimating its cost and environmental impacts because the
project should be based on comprehensive analyses of many relevant factors,
such as land availability, earthwork, maintenance, life-cycle cost, demand,
land-use, user travel time, environmentally sensitive areas, safety, effects
on the performance of other transportation modes, and effects on regional
development.

A survey of the procedures adopted by highway agencies [e.g., United
States Department of Transportation (USDOT)] reveals that a combination
of engineering judgment, an environmental impact analysis, and manual
cost-benefit evaluation is generally followed for assessing various alternatives for new highway construction, realignment, or expansion. The alternatives are ranked through a weighed-criteria analysis in which a set of viable
alternatives are manually identified and ranked based on a total score. This
score consists of weights assigned for various criteria, such as economic and
environmental impacts, cost, and safety. The alternatives are then presented
to the stakeholders in town meetings, in which public support is sought.
The alternative that wins the maximum support is usually chosen for the

actual construction, and proceeds to the detailed design stage. A preliminary estimate of total cost for planning, design, right-of-way acquisition, environmental impact mitigation, and construction is then prepared using historical unit cost data, engineering judgment, and through trial and error.

1.2   Highway Alignment Optimization Problems

Adding a new highway to an existing road network or widening an existing road may be considered for improving the performance of that road network. Such a supply action increases capacity of the road network, and highway users may thus save travel time and vehicle operation costs. Positive economic impacts may also result from the transportation system improvement. Nevertheless, finding the location and alignment profile of a new highway that best improves the existing road network is a complex problem since many factors affect road construction. Changes in network traffic flow patterns from the new highway addition, various costs associated with highway construction, design specifications, safety, environmental impacts, and political issues affect the project. In addition, profiles and impacts of the new highway may depend significantly on how and where it is connected to the existing road network.

For new highway construction, an optimized road connecting specified points or sections on existing highways is desired. In expansion projects, an optimized road between tightly specified bounds may be desired. The determination of the best available option for new highway construction, realignment, or expansion falls in the general area of *Highway Location and Alignment Optimization*, and it is defined as

> *"an optimization process to find location and alignments of an economical path connecting existing roads that minimizes total costs, including agency, user, environmental, and socio-economical costs in the network, while satisfying geometric design, operational, environmental, and geographical constraints."*

In a conventional highway design process, highway planners and engineers identify relatively few candidate alignments, and then narrow their focus to the detailed alignment design. Many possible alternatives which are much better than the one selected for final design may exist but get no consideration, as Figures 1.1 and 1.2 suggest. Considerable time and cost is needed to find the best one among the candidate alignments since the

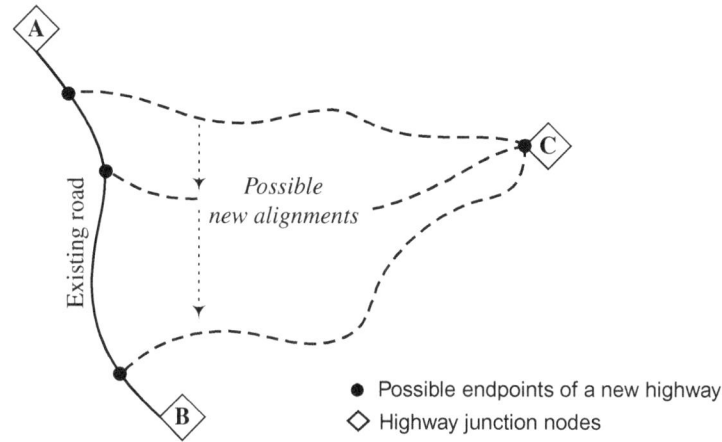

(a) Case 1: Either start or end point of a new alignment is undefined

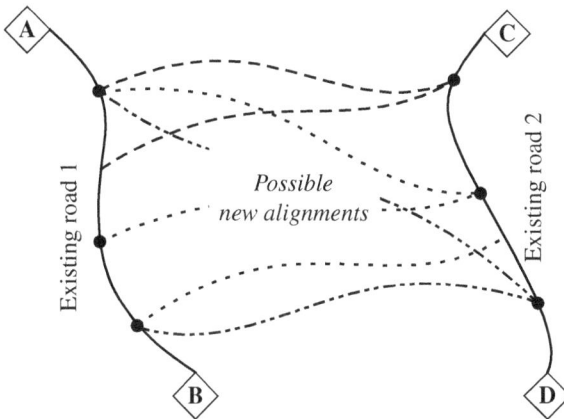

(b) Case 2: Both endpoints of a new alignment are undefined

Figure 1.1 Possible highway alignments connecting existing roads

conventional approach requires repetitive manual processes for performing detailed design and evaluating all the alignments. It is often the case that many critical issues, including unusual topographic and geographic issues as well as design requirements arise at later stages, inflating the preliminary cost estimates of the final alternative. Seeking additional funding at later stages often delays the scheduled completion of the project since funds are typically approved on an annual basis and are too scarce to support all

Possible endpoints of a new highway

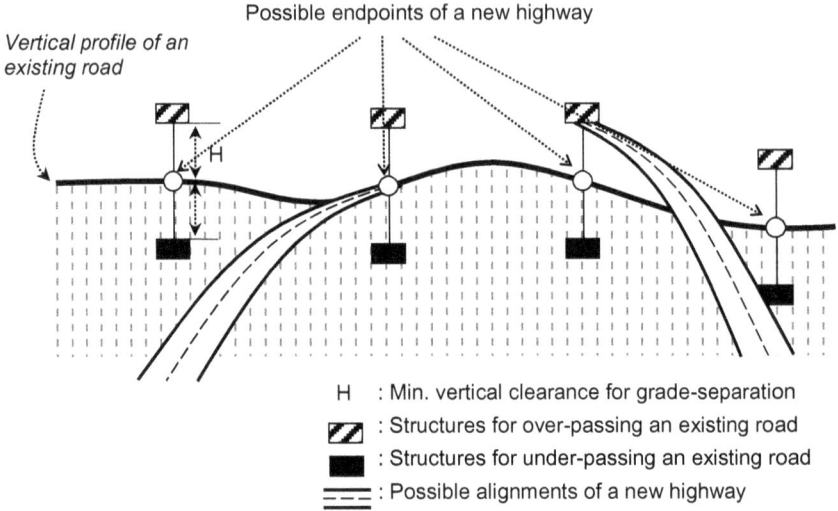

H   : Min. vertical clearance for grade-separation

▨   : Structures for over-passing an existing road

■   : Structures for under-passing an existing road

=== : Possible alignments of a new highway

Figure 1.2 Possible highway endpoints along an existing road

worthwhile projects. Thus, an intelligent computation tool that finds the best highway alignment for given requirements, and which can be easily re-run for changing objectives, constraints or preferences, can save billions of dollars in resources and life-cycle costs of capital, operation, and user time.

To overcome complex and time-consuming limitations in the traditional highway location selection and alignment design process, many studies have proposed automated and computerized optimization methods, such as Dynamic Programming, Sequential Quadratic Programming, Network Optimization, Particle Swarm Optimization, and Genetic Algorithms. However, some of these methods, although performing well in certain aspects, still have considerable weaknesses and are not widely utilized in real world applications. They may require special data formats, consider only limited number of factors associated with highway construction, or provide unrealistic highway alignments for real world applications with strong underlying assumptions; some models, despite providing relatively good solution alignments, may have significant computation burdens due to inefficient solution search methods.

Suppose that construction of a new highway is considered to improve the traffic performance of an existing highway network (see Figure 1.3).

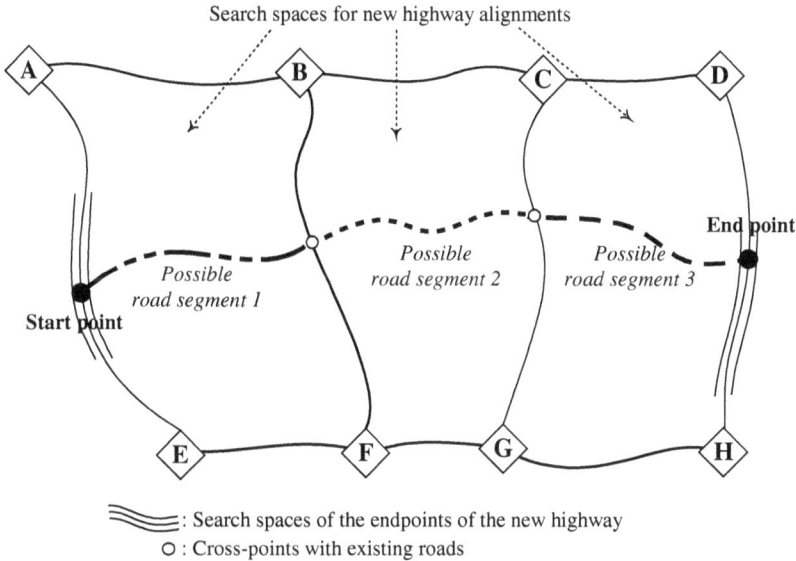

Figure 1.3 An example road network with addition of a new highway consisting of multiple possible road segments

Then, highway planners and designers will try to find an economical path that minimizes the total construction cost as well as improves the traffic performance of the network, while satisfying geometric design, operational, and geographical constraints (including user preferences[1]). Horizontal and vertical profiles of a new highway may significantly vary depending on the locations of its endpoints (i.e., where to connect a new highway in the existing road network) as well as factors associated with its construction (such as topography and land-use of the study area and its design standards). Changes in traffic flow patterns of the road network may also vary depending on where the alignments are connected on that network and their total distances. Prior to this, no models have been found that jointly consider these issues.

This monograph takes such considerations into account in optimizing highway alignments, and provides a comprehensive overview of the optimization problem with detailed review of various methods for searching

---

[1] User preferences may include preferences of highway planners and designers or opinions from public hearing that affect right-of-way of a new highway.

the best location and alignment design of a highway. Key topics covered include, but are not limited to:

- Genetic Algorithms (GAs) and other Artificial Intelligence (AI) techniques for optimizing highway location selection and alignment design;
- Modeling three-dimensional (3D) highway alignments;
- Costs associated with highway construction, operations, and maintenance;
- Highway geometric design, traffic, socio-economic, and environmental considerations in highway alignment optimization;
- Bi-level highway alignment optimization; and
- Applicability of the GAs-based highway alignment optimization models with case studies.

1.3   Organization of the Monograph

This monograph is organized into four parts. PART I consists of Chapters 1 to 3, and overviews highway location generation, selection, and alignment optimization problems. Relevant cost items and constraints associated with highway location selection and alignment design are discussed. Existing optimization models that employed AI techniques to solve the problem are also reviewed.

PART II starts with Chapter 4 where a highway alignment optimization method based on AI called the Highway Alignment Optimization (HAO) model is discussed. The HAO model uses GAs, a powerful AI method, to generate, evaluate, and optimize highway location and alignment design in a 3D space. Highway Alignment Optimization model formulation with relevant highway costs and constraints is discussed in Chapter 5. How Geographic Information System (GIS) and the GA-based HAO model are communicated during the optimization process are also discussed in this chapter. Custom constraint handling techniques developed to improve computational efficiency and solution quality of the GA-based HAO model are discussed in Chapters 6 to 8.

In PART III (Chapters 9, 10, and 11), the highway location and alignment optimization problem is extended to include various user cost components which account for traffic effects of a new highway on an existing road network. For that, the problem is reformulated as a bi-level optimization problem where the highway alignment optimization problem is treated as an upper level problem, while a traffic assignment problem is treated as a lower level problem. In this bi-level optimization framework, the alignments of a new highway as well as its junction points with existing roads are simultaneously optimized.

The GA-based HAO model is applied to several example projects in PART IV (Chapters 12, 13, and 14) to investigate the model applicability and usability to real-world problems. A series of analyses of sensitivity to key model parameters is also conducted along with the case study to demonstrate model capability. Each of the chapters concludes with optimized highway alignments and a detailed cost breakdown. Highway planners and designers may benefit from this model, which offers well optimized candidate alternatives, developed with automated GIS data extraction and GA-based comprehensive alignment optimization, rather than merely satisfactory alternatives in the planning stages of new highways. PART IV concludes with Chapter 15 where potential improvements of the HAO model and other exciting developments of highway location search and alignment optimization applications are discussed.

Chapter 2

Highway Cost and Constraints

This chapter investigates major cost items that should be considered in highway location selection and alignment development. Such an investigation is essential since these are criteria that are most commonly considered in evaluating highway alignment alternatives.

Many cost components directly or indirectly affect the construction of new highways. Besides the initial construction costs, which are directly counted for highway construction (e.g., earthwork, land acquisition, pavement and drainage costs), user costs and environmental impact should also be considered. It is important to note that all dominating and alignment-sensitive costs should be considered and precisely formulated for a good highway optimization model; dominating costs are those which make up significant fractions of the total cost of a new highway alignment, and alignment-sensitive costs are those which vary significantly with relatively slight changes in alignment geometries.

Normally, highway user costs (such as travel time cost and vehicle operating cost) are the most dominating ones as they persist over the entire design lifetime of the highway and the users' value of time is usually higher than other costs associated with highway construction. Structure costs (e.g., bridges and interchanges costs) and earthwork costs may dominate if a highway is constructed in a mountainous area. A highway passing through an urban area may have a high percentage of right-of-way cost, since the required land acquisition cost of that area may be relatively higher than other costs.

A number of studies (Winfrey, 1968; Moavenzadeh *et al.*, 1973; OECD, 1973; Wright, 1996; Jong, 1998; Jha, 2000; Kang, 2008) have discussed the costs associated with highway location selection and alignment design,

Table 2.1 Classification of major highway costs and impacts

Classification		Examples
Construction costs		Earthwork, pavement, right-of-way, structures
Maintenance costs		Pavement resurfacing, bridge/tunnel operation and maintenance, lighting
	Vehicle operating cost	Fuel, tire wear, depreciation of vehicles
User costs	Travel time costs	Vehicle hours × unit value of time
	Accident costs	Predicted numbers of accidents and their costs
Environmental and socio-economic impacts		Loss of environmentally sensitive areas, loss of socio-economic areas, noise, air pollution

and identified major costs that should be considered for optimizing highway alignments. The major costs are summarized in Table 2.1. Note that costs required in the highway planning and design stages (such as consulting and data collection costs) are not included here among the major costs since they may not be sensitive to alignments of highway alternatives.

2.1 Initial Construction Costs

The construction costs are the major agency costs that directly affect highway agencies (e.g., local and federal governments) or highway construction companies. Normally, costs required for earthwork, pavement, right-of-way, structures (e.g., bridges and interchanges), and miscellaneous items (such as fencing and guardrails) are included in this category. Jong (1998) and Jha (2000) reclassified them into four sub-categories based on the characteristics of each cost component: (a) volume-dependent cost, (b) length-dependent cost, (c) location-dependent cost, and (d) structure cost. Such a classification is quite useful for quantifying the construction costs and representing them in the alignment optimization process.

The earthwork cost is a volume-dependent cost since it is quantifiable based on the amount of earthwork volume required for highway construction. Some unit costs related to the earthwork (such as unit embankment and excavation costs) are needed to estimate that cost. The right-of-way cost, which includes land acquisition costs and property damage and compensation costs, can be classified as the location-dependent costs. The length-dependent cost is defined as the cost proportional to alignment length. Pavement cost and road superstructure and substructure costs (such as fencing, guardrails, and drainage costs) can be included in this category. In highway engineering, structures normally include bridges, tunnels, interchanges, intersections, and overpasses or underpasses. Costs required for building those structures belong to the structure cost category. All costs that are dominating and sensitive to the alignment should be included in the alignment optimization process.

2.2 Highway Maintenance Costs

The maintenance costs occur throughout the design life of the road. Therefore, for life-cycle cost estimates, these costs are generally discounted over the road's life at an appropriate interest rate. The maintenance costs include costs for routine maintenance and operation of bridges and tunnels as well as preventive maintenance costs (such as costs required for repairing pavements, guardrails, and medians) and road rehabilitation costs of basic road segments.

2.3 User Costs

The highway user costs are sometimes also called traffic costs and usually include travel time, vehicle operating, and accident costs. These costs also occur throughout the design life of the road, and thus should be estimated as a life-cycle cost. In a highway improvement project, these costs are normally used for a user benefit analysis, by comparing their values estimated before and after the project. The travel time cost can be computed with the users' travel time estimated in a certain condition (e.g., specific time and scenario (with or without a new highway)) of a highway network and their value of time estimated externally. The vehicle operating costs typically include estimated fuel consumption and vehicle depreciation costs. The accident costs are usually estimated with unit accident cost and accident rates predicted from an accident regression analysis.

Note that the user cost items are the dominating costs, and they are sensitive to alignment length as well as to the locations where a new alignment is connected to existing road networks. Therefore, the user cost should also be considered in the optimization process. The methods for estimating these costs are well discussed in the American Association of State Highway and Transportation Officials (AASHTO) manual for "User and Non-user Benefit Analysis for Highways" (2010).

2.4   Environmental and Socio-economic Impacts

Construction of a new highway may also significantly affect environmentally sensitive areas (such as wetlands and historic areas) and human activities of the existing land-use system, and even cause air pollution and increased noise level. These impacts are often the most important issues in the modern highway construction projects; hence, they should also be accounted for in the alignment optimization process.

Jong *et al.* (2000) and Kang *et al.* (2012) considered the environmental and socio-economic impacts of highway alternatives in the alignment optimization problem by using a penalty concept; they assign high penalties to the areas considered as environmentally sensitive or "no-go" regions. However, it should be noted that a detailed trade-off analysis (or a decision-making process) may be required to use the penalty concept if a highway project is complex so that there are different levels of importance in the environmentally and/or socio-economically sensitive regions. An example of the trade-off analysis for a real highway construction project is described in Chapters 12 and 13.

Jha (2000) provides more detailed discussions for the environmental issues associated with the new highway construction. He comprehensively formulates highway environmental costs in the alignment optimization process with a GIS-based application. Kim (2001) considers the noise and air-pollution effects of the new highway alternatives in the optimization process although they are not significantly sensitive to highway alignments. A previously developed noise model (Haling and Cohen, 1996) and air-pollution cost model (Halvorsen and Ruby, 1981) are adopted to represent the noise and air-pollution effects on solution selection.

Costs for highway planning and alignment design may be neglected in alignment optimization problems because they are insensitive to various highway alternatives. Furthermore, their impacts on the total cost of a highway construction project are not significant (i.e., not dominating).

2.5 Highway Constraints

Normally two types of constraints are considered in highway location selection and design. These include (i) design constraints and (ii) environmental and geographical constraints. The former constraints are usually based on recommended design standards (e.g., AASHTO, 2011); the latter ones are sensitive to many complex factors associated with topology, land-use of the project area, and even preferences of decision-makers.

2.5.1 Geometric Constraints

Basically, the geometric design of a highway determines the horizontal alignment, vertical alignment, and cross-section of that highway. The horizontal alignment of a highway, which is a projection of a 3D highway onto a two-dimensional horizontal space (i.e., XY coordinate system), generally consists of three types of design elements: tangent segments, circular, and transition curves. On the other hand, the vertical alignment is a projection of a design line on a vertical plane as if all horizontal curves were stretched to straight, and composed of a series of graded-tangents joined to each other by parabolic curves. According to AASHTO (2011), the most important design constraints required for constructing the horizontal and vertical alignments are:

- Minimum horizontal curve radius with a side friction factor based on superelevation
- Sight distance on horizontal curves [i.e., horizontal sight offset (HSO)]
- Minimum superelevation runoff lengths and/or spiral transition curve lengths (only if transition curves are considered as a part of the horizontal curved section)
- Maximum gradients of vertical alignments

- Sight distance on crest and sag vertical curves (i.e., minimum length of crest and sag vertical curves)
- Minimum vertical clearance for highway crossing and bridge construction

2.5.2   Geographical and Environmental Constraints

Besides the design constraints stated earlier, geographical and environmental constraints should also be considered in the highway design process. Nowadays, these constraints are considered among the most important criteria in actual highway construction projects. For example, in the United States, all roadway projects that involve federal funding must have National Environmental Policy Act (NEPA) approvals unless it can be demonstrated that no permit is required (Eccleston, 2008). NEPA's main policy is to guarantee that the government provides proper consideration to the environment prior to undertaking any major federal action that significantly affects the environment.

Several types of land-use that may affect most highway projects are listed in the following. These can be modeled and treated as geographical and/or environmental constraints when optimizing highway location and alignments. As such, a highway alternative that has the least amount of negative impact and the largest amount of positive impacts as a whole on its surrounding environment is found through an optimization process.

- Environmentally sensitive areas (e.g., wetlands and historic districts)
- Socio-economically sensitive areas (e.g. residential and commercial areas)
- Control areas defined by highway decision makers (e.g., political issued areas)
- Fixed points (or areas) constraints through which the highway alignment must pass

Chapter 3

Review of Artificial Intelligence-based Models for Optimizing Highway Location and Alignment Design

Artificial intelligence (AI) is a field of study that synthesizes and analyzes computational agents that act intelligently (Polle and Mackworth, 2017). AI, however, is normally characterized as computer applications in engineering fields that show features similar to intelligence of living creatures, such as evolution, collaborative behavior, processing of natural language, dealing with spatial structures, logical reasoning with fuzzy concepts, and learning (Transport Research Board (TRB), 2012). Therefore, heuristic and metaheuristic methods, designed for solving complex problems through a process of learning and discovery, are considered as a core field of AI in engineering. Several AI methods have been applied in highway location and alignment optimization problems. These include (i) Genetic Algorithms (GAs), (ii) Swam Intelligence (SW), and (iii) Simulated Annealing (SA). This chapter discusses computer applications or models that used those AI methods to solve highway location and alignment optimization problems.

3.1 Genetic Algorithms-based Optimization Models

Genetic algorithms (GAs), derived from Darwin's theory on genetic evolution, are an AI method that has been popularly employed for solving complex optimization problems that cannot easily be solved by traditional analytical approaches. GAs are adaptive search methods based on the principles of natural evolution and survival of the fittest. Among many different methods proposed for highway location selection and alignment optimization, GAs have been the most widely-investigated due to their ability to realistically handle the complex fitness functions featuring nonlinearities and discontinuities while including mechanisms to escape local optima (Davey *et al.*, 2017). GAs with effective constraint handling methods can avoid getting trapped in local optima and yield near optimal solutions in a continuous search space, by providing a pool-based search rather than single solution comparison as in other heuristics. In practice, this AI method has proven to be simple, fast, and effective. Furthermore, little special knowledge is needed to apply it, and users can apply the optimization technique to problems where a proof of global optimality is not necessary (TRB, 2012). The effectiveness of GAs in highway location selection and alignment optimization can be described in terms of the following key advantages:

- Allows a continuous search space
- May be customized with special operators based on problem structure to efficiently solve large and complex problems
- Can find globally or near globally optimal solutions
- Can consider most relevant highway costs and constraints
- Can yield realistic alignments

3.1.1 Highway Alignment Optimization (HAO) Models by the UMD Research Team

The highway alignment optimization (HAO) model developed by the University of Maryland (UMD) research team is probably the most widely known approach that exploits GAs for optimizing highway alignments. It is designed to minimize costs associated highway users, highway agencies, and environmental impacts, subject to the geometric design, environmental and geographical constraints. It is called the HAO model, and has been extensively refined by the UMD research team in various ways to deal with the complex highway location selection and alignment optimization problems. The major contributors to the model include: Paul Schonfeld, Jyh-Cherng Jong, Manoj Jha, Eungcheol Kim, and Min-Wook Kang. Table 3.1 shows a sequence of the model development with a list of publications produced by the team members.

The first version of the model (i.e., HAO 1.0) employed GAs with several specialized genetic operators to jointly optimize horizontal and vertical alignments (Jong, 1998; Jong and Schonfeld, 2003). In the model, the highway alignments were represented in a three-dimensional (3D) continuous

Table 3.1 HAO model development history and contribution

Version	HAO model development studies	Main contribution
HAO 1.0	• Jong, J.-C. (1998), "Optimizing Highway Alignments with Genetic Algorithms", Ph.D. Dissertation, University of Maryland, College Park. • Jong, J.-C., Jha, M.K., and Schonfeld, P. (2000), "Preliminary Highway Design with Genetic Algorithms and Geographic Information Systems", *Computer-Aided Civil and Infrastructure Engineering*, 15(4), pp. 261–271. • Jong, J.-C. and Schonfeld, P. (2003), "An Evolutionary Model for Simultaneously Optimizing 3-Dimensional Highway Alignments", *Transportation Research*, Part B, 37(2), pp. 107–128.	• Use GAs for highway alignment optimization • Develop customized genetic operators for optimizing highway alignments • Jointly optimize horizontal and vertical alignments • Develop basic highway cost functions to evaluate initial construction cost of alternative alignments
HAO 2.0	• Jha, M. K. and Schonfeld, P. (2000a), "Integrating Genetic Algorithms and GIS to Optimize Highway Alignments", Transportation Research Record No. 1719, pp. 233–240. • Jha, M. K. and Schonfeld P. (2004), "A Highway Alignment Optimization Model using Geographic Information Systems", *Transportation Research*, Part A, 38(6), pp. 455–481.	• Extend **HAO 1.0** by integrating a GIS analysis and GAs-based optimization. As such, the model can realistically evaluate right-of-way costs of alternative alignments.

(*Continued*)

Table 3.1   *(Continued)*

HAO 3.0	• Kim, E., Jha, M. K., Lovell, D. J., and Schonfeld, P. (2004b), "Intersection Cost Modeling for Highway Alignment Optimization", *Computer-Aided Civil and Infrastructure Engineering*, 19(2), pp. 136–146. • Kim, E., Jha, M. K., Schonfeld, P., and Kim, H. (2007), "Highway Alignment Optimization Incorporating Bridges and Tunnels", *Journal of Transportation Engineering*, 133(2), pp. 71–81.	• Develop cost functions for major highway structures. • Extend **HAO 1.0** by incorporating the structure cost functions into the model objective function.
HAO 4.0	• Kang, M. W., Schonfeld, P., and Jong, J.-C. (2007), "Highway Alignment Optimization through Feasible Gates", *Journal of Advanced Transportation*, 41(2), pp. 115–144. • Kang, M.-W., Schonfeld, P., and Yang, N. (2009), "Prescreening and Repairing in a Genetic Algorithm for Highway Alignment Optimization", *Computer-Aided Civil and Infrastructure Engineering*, 24(2), pp. 109–119. • Kang, M.-W., Jha, M.K., and Schonfeld, P. (2012), "Applicability of Highway Alignment Optimization Models", *Transportation Research*, Part C,21 (1), pp. 257–286.	• Develop two direct constrain handling methods (called "Feasible Gate" and "Prescreening & Repairing" methods) for effective GAs-GIS based alignment optimization. • Improve the computation efficiency and solution quality of **HAO 3.0** by incorporating the direct constrain handling methods into the GAs-based optimization process. • Apply the **HAO 4.0** to actual highway projects to demonstrate its applicability to real-world problems

Table 3.1 *(Continued)*

HAO 5.0	• Kang, M.-W. Shariat, S., and Jha, M.K. (2013), "New Highway Geometric Design Methods for Minimizing Vehicular Fuel Consumption and Improving Safety", *Transportation Research*, Part C, 31, pp. 99–111.	• Develop a sight distance module for design consistency check and safety evaluation • Develop a vehicle dynamic module to estimate vehicle operating costs for varying highway geometries
HAO 6.0	• Yang, N., Kang, M.-W., Schonfeld, P., and Jha, M.K. (2014), "Multi-objective Highway Alignment Optimization Incorporating Preference Information", *Transportation Research*, Part C, 40, pp. 36–48.	• Develop and apply a Hybrid Multi-Objective Genetic Algorithm (HMOGA) in the alignment optimization process to 1) incorporate highway designers' prior knowledge in the model objective functions, and 2) speed up the convergence towards possibly preferred solutions
HAO 7.0	• Kang, M.-W., Yang, N., Schonfeld, P., and Jha, M.K. (2010), "Bilevel Highway Route Optimization", *Transportation Research Record: Journal of the Transportation Research Board*, No. 2197, pp. 107–117.	• Reformulate the HAO problem as a bi-level optimization framework to determine a new highway location that best improves an existing roadway network as well as to optimize its alignments based on geometric, user and agency costs, and operational considerations

space, while considering major highway design elements (e.g., tangents and circular curves for horizontal alignments; gradients and parabolic curves for vertical alignments). Several highway cost functions (e.g., earthwork and length-dependent costs) were also developed for use in the model objective function. XYZ coordinates of a series of points of intersections (PI's), which outline the shape and location of a highway alignment, were used as decision variables in the HAO model, and they were directly or indirectly formulated to estimate highway costs.

In the second version of the model (i.e., HAO 2.0), a GIS analysis was added to realistically reflect a real world problem. In HAO 2.0, the GIS is primarily used for right-of-way (ROW) cost estimation, while the GAs are used for generation, evaluation, and optimization of highway alignments (Jha and Schonfeld, 2000a, 2000b and 2004). In HAO 3.0, Kim (2001) and Kim *et al.* (2004a and 2004b) further extended HAO 1.0, by incorporating major structure costs in the model objective function. As such, HAO 3.0 can determine (during the alignment optimization process) whether and where bridges or tunnels are preferable to embankments or deep cuts, respectively.

In HAO 4.0, two direct constraint handling methods (called "Feasible Gate" and "Prescreening & Repairing" methods) were developed to enhance the computation efficiency and solution quality of the previously developed models (Kang *et al.*, 2007 and 2009). These methods were intended to avoid generating infeasible solutions that are outside the acceptable bounds and thus to focus the search on the feasible solutions. To show the applicability of HAO 4.0 to a real-world problem, two actual highway projects in Maryland, USA, were analyzed using the model. The analysis results show that the alignments optimized by the model are quite similar to those obtained through conventional manual methods by a state agency, but the model can greatly reduce the time required for highway planning and design as well as produce much less expensive solutions.

The model was further improved in HAO 5.0 to include i) a sight distance module for design consistency check and safety evaluation and ii) a vehicle dynamic module to estimate vehicle operating costs for varying highway geometries (Kang *et al.*, 2013). Such an integrated modeling framework helped the model to search for green and environmentally sustainable highways. A Hybrid Multi-Objective Genetic Algorithm (HMOGA) was developed and applied in HAO 6.0 to account for preferences of different stakeholders to a new highway development project. Such a multi-objective feature allowed the model to examine the trade-offs among objectives that represent the possibly conflicting preferences of transportation agencies, road users, and the affected society.

In HAO 7.0, the highway alignment optimization problem was reformulated as a bi-level program not only to optimize highway alignments, but also determine their best location within a small road network. In it the variation of traffic impacts on the existing road network caused by the addition of different highway alignments and locations were assessed in the optimization process. Readers may refer to Parts II and III of this monograph for details of the HAO model, including model structure, optimization methods, input, output, and capabilities.

It may also be noted that the GA-based alignment optimization concepts developed for the HAO model were applied by Lai (2012) and by Lai and Schonfeld (2012 and 2016) in optimizing rail transit alignments, including station locations along the rail transit lines. Those optimization methods considered how vertical alignments could be designed to reduce vehicle energy use, when and how 3D rail transit alignments should follow existing surface streets, (e.g., using cut & cover construction), and the effects of station locations on demand and passenger access paths to stations.

3.1.2   Other Models that Use Genetic Algorithms for Optimizing Highway Alignments

There are several studies in the literature that used GAs for optimizing highway alignments, besides the HAO model developed by the UMD research team. These include, but are not limited to (i) *Optimal Vertical Alignment Analysis for Highway Design* by Fwa *et al.* (2002); (ii) *Using Genetic Algorithms for Optimizing the PPC in the Highway Horizontal Alignment Design* by Bosurgi *et al.* (2014); (iii) *Optimal Road Design through Ecologically Sensitive Areas Considering Animal Migration Dynamics* by Davey *et al.* (2017); (iv) *Forest Road Profile Optimization Using Meta-heuristic Techniques* by Babapour *et al.* (2018).

Fwa *et al.* (2002) optimized vertical alignments with the assumption that horizontal alignments are initially given. In their model, GAs with a constant static penalty function were employed to handle infeasible solutions. A huge constant value (which was pre-specified as a model input) was added to the model objective function whenever the solution violates design constraints regardless of severity of the violation. It should be noted, however, that such a constant penalty function is generally inferior to a soft penalty function which adds more severe penalty with distance from the feasibility condition (Goldberg 1989; Richardson *et al.* 1989; Smith and Coit, 1997). A huge constant value may cause serious errors by leading to large unsmooth steps during the optimization process, and thus may often fail to obtain optimal solutions.

Bosurgi *et al.* (2014) also used GAs to optimize polynomial parametric curves (PPCs) in the highway horizontal alignment design. They developed several GA operators to generate PPC parameters, which change the curve shape and length. Thus, various PPCs are generated, relocated, and evaluated through the alignment optimization process. This research showed that GAs provide a good solution that guarantees compliance with the design prescriptions.

Davey *et al.* (2017) developed a useful framework for highway alignment optimization which considered the effects of road construction on animal movement and mortality. This study incorporated an animal movement and mortality model into the GA-based HAO model, developed by the UMD research team, to find the optimal alignment through an ecologically sensitive region. The study also provided an excellent summary of previous literature associated with highway alignment optimization and empirical ecological models.

Babapour *et al.* (2018) compared a GA and a Particle Swarm Optimization (PSO) algorithm to find a near optimal forest road profile on a mountainous terrain. Results showed that the GA and PSO algorithm both could reduce earth work costs, as compared to a manual design method, while providing optimized vertical alignments that satisfy design constraints. Results also showed that the GA was superior to the PSO algorithm in terms of computation time and costs, particularly for the forest road profile optimization problem where a great number of decision variables are required.

3.2   Other Artificial Intelligence-based Optimization Models

3.2.1   Swarm Intelligence-based Models

Swarm intelligence (SI) is a branch of AI, based on the study of individual's behavior in various decentralized systems (Teodorovic, 2008). SI employs intelligence of collective behavior from a large group of species to make decisions. Among several SI-based algorithms available in the literature, particle swarm optimization (PSO) algorithms have been popularly used by transportation researchers in optimizing highway alignments (Shafahi and Bagherian, 2013; Bosurgi *et al.*, 2013 and 2017). PSO is a powerful tool inspired by the social behavior of bird flocking and fish schooling (Kennedy and Eberhart, 1995). Similar to GAs, PSO is an evolutionary computation tool and a metaheuristic, which can be applied for a variety of optimization problems with few or no assumptions about the problems.

PSO performs a population-based search, using a swarm of particles to represent potential solutions within a search space. The fitness of each particle represents the quality of its position, and each particle is characterized by its position, velocity, and a record of its past performance (Shafahi and Bagherian, 2013). The particles move over the search space with a certain velocity, which determines direction and speed of particle's movement. The velocity of each particle is influenced by best solutions that were found by its neighbors and/or globally as well as its local best solution (i.e., position) found so far. The swarm will converge to optimal positions eventually (Engelbrecht, 2007; Bonabeau *et al.*, 1999).

Several studies proposed PSO algorithms to optimize highway alignments (Shafahi and Bagherian, 2013; Bosurgi *et al.*, 2013 and 2017). They develop custom PSO algorithms to find highway alignments in a 3D search space that minimize the sum of several highway cost items. Coordinates of PI's were used as decision variables of the models. Several operators, motivated by the UMD research team's GAs-based HAO models (Jong *et al.*, 2000; Kang *et al.*, 2012) were also proposed in the PSO models to improve the efficiency of a solution search process. It is important to noted that all these PSO models adopted the core concepts of the HAO model, except solution search algorithms. Decision variables (i.e., xyz coordinates of PI's), modeling of cost functions with decision variables, optimization framework, objective functions, constraints, and a GIS analysis were all adopted from the UMD team's HAO model (Jong *et al.*, 2000; Jha and Schonfeld, 2004; Kang *et al.*, 2007b, 2009, and 2012). The efficiency of the PSO algorithms was verified with several numerical examples. The studies showed that the PSO algorithms could be used as an alternative of GAs for optimizing highway alignments. Bosurgi *et al.* (2017) later included Polynomial Parametric Curve (PPC) in horizontal curve design when optimizing highway alignments using a PSO algorithm. The study showed that the adoption of the PPC would allow higher levels of comfort for road users than traditional transition curves.

Vazquez-Mendez *et al.* (2018) developed a geometric model that optimizes alignments of a road joining two terminals. In this study, the alignment optimization problem is formulated as a smooth constrained optimization problem, and transition curves were included in the horizontal alignment design. Clothoids are included as transition between tangents and circular curves in the horizontal alignment. The study also used a two stage optimization process; first use a PSO algorithm to find an initial set of solution alignments, and then apply the initial solution to a Sequential Quadratic Programming (SQP) algorithm to optimize horizontal and vertical alignments. Numerical results were presented in the study to demonstrate the applicability of the model.

PSO-based methods for alignment optimization have recently been developed for railway applications, for instance by Hasany and Shafahi (2017). Railway alignment optimization applications are similar in many ways to highway applications, but may require some different constraints. In particular, Pu *et al.* (2019a) combine a PSO algorithm with genetic operators to optimize mountain railway alignments, which is a relatively difficult problem. Ghoreishi *et al.* (2019) use a PSO algorithm for optimizing railroad alignments, with special emphasis on incorporating bridges and tunnels in the alignments.

3.2.2 Neighborhood Search Heuristic

Cheng and Lee (2006) also proposed 3D alignment optimization model with a neighborhood search-heuristic for finding horizontal alignments and a mixed integer programming (MIP) method for finding vertical alignments. Several cost components (such as earthwork costs and bridge and tunnel costs) are included in the model objective function. The key contribution of this paper is that transition curves are used to realistically represent the curved sections of horizontal alignments while considering various design constraints associated with the curves. Besides the alignments profile (horizontal and vertical), a speed profile for heavy vehicles operating on the resulting alignments is created as a model output. Despite the contribution, several limitations are found in the model. First (i) Cheng and Lee's (2006) model finds 3D highway alignments with a two-stage (conditional) approach; it sub-optimizes a horizontal alignment first, and then sub-optimizes the vertical alignment based on the horizontal alignment created. It should be noted, however, that optimizing horizontal and vertical alignments (i.e., 3D alignments) simultaneously is clearly preferable to sub-optimizing the vertical alignment for a previously sub-optimized horizontal alignment since such a conditional optimizing process is less likely to avoid local optima. In addition, (ii) alignment right-of-way cost (which includes land acquisition and property damage costs), a clearly dominating and sensitive cost in highway construction project, is not considered in the model. Note that it would be desirable that a model for optimizing highway alignment should directly exploit a GIS database because most spatial information is becoming available in such computer-readable form. This includes realistic shapes of land parcels, property values, and even various land-use patterns. Finally, in Cheng and Lee's (2006) model (iii) the untouchable area (i.e. no-go area of alignments) was assumed to be circular in shape although in reality it could have any shape.

3.3   Other Methods Tried with AIs for Optimizing Highway Alignments

3.3.1   Distance Transform

The distance transform (DT) method has mainly been used recently for railway alignment applications. Its main proponents are a research group at Central South University in Changsha, China, led by Prof. Hao Pu. Their published works using the DT approach include Li *et al.* (2016), Li *et al.* (2017), Pu *et al.* (2019b), and Pu *et al.* (2019c). This team has extended the DT approach to 3D railway alignment optimization and is currently working on distributing the algorithm among parallel processors to speed up computations (Song *et al.*, 2020). The success of the DT approach in optimizing relatively complex mountain railway alignments suggests that this approach may also be quite promising for optimizing highway alignments in complex topographic conditions.

Kang *et al.* (2011) also used the DT approach to find best paths for a military path planning application. They modeled critical factors (such as terrain slope and elevation, surface travel distance, degree of bumpiness in various land-uses, known and unknown enemy locations, and exposure to enemies) that affect the path planning in a combat environment, and used a DT algorithm to find the best paths of unmanned ground robots (UGRs) between any points. Various GIS layers were employed to build a weighted travel cost map, which was eventually used to find the least-cost path of the UGR through a DT process. This application suggests that the DT method has great potential for testing and evaluating various missions to be performed by UGRs before their deployment in a real-world situation.

3.3.2   Multi-objective Optimization

Several researchers have sought to apply multi-objective optimization concepts to highway alignments. The earliest such works appear to be by Jha and Maji (2007), Maji and Jha (2009, 2011 and 2012), and by Yang *et al.* (2014). Such multi-objective formulations are of special interest when various objectives are subject to different value judgments or are difficult to measure in commensurate units, such as construction costs, environmental impacts, and expected fatalities. In such cases, decision-makers may prefer to consider tradeoffs among multiple alternatives along a Pareto Front (Pareto, 1906) rather than rely on a single weighted objective function. Thus, Yang *et al.* (2014) formulates the highway alignment optimization problem with two objectives, namely (1) agency costs and (2) effects on environmentally sensitive areas. They show how the preferences of decision-makers can be reflected in the alignment selection. Hirpa *et al.* (2016) focus on comparing several mathematical approaches for bi-objective alignment optimization. These are demonstrated on a simplified problem in which only the earthwork cost and the alignment length are treated as the objectives to be optimized.

3.3.3   Multiple Path Optimization

Some researchers have considered the problem of identifying relatively few near-optimal alignments that differ significantly in their locations. Thus, Jha (2000) considered how such significantly different alignments could be generated and how their dissimilarity could be measured. Pushak *et al.* (2016) compared five algorithms for solving the dissimilar multipath problem and found that a bidirectional search algorithm yielded the fastest and best results.

3.3.4   Other Objectives Considered

It may be noted that some researchers focused on some limited objectives for alignment optimization, unlike the very comprehensive objective functions sought in the GA-based HAO models by the UMD research team. For example, Easa and Mehmood (2008) focused on the safety aspects of horizontal alignments. Hare *et al.* (2015) focused on optimizing the earthwork costs associated with a road's vertical alignment using a mixed-integer linear programming approach. Hirpa *et al.* (2016) focused mainly on earthwork cost and road length. Kim *et al.*, (2019) optimized vertical alignments for rail transit lines by considering construction costs as well as the operating costs of trains at optimized speeds.

Jha and Kang (2009) developed a GIS-based model that automatically detects highway segments where installation of noise barriers may be warranted. This model calculates the required number of noise walls, their lengths, and locations along an existing or a new highway alignment being developed. This model can be incorporated into the GA-based HAO model to make the highway location and alignment optimization process more effective.

Kang *et al.* (2013) focused on fuel consumption and safety aspects of horizontal and vertical alignments. They developed fuel consumption models to account for variability of fuel consumption by vehicles depending on variability in highway alignment geometry. A sight distance model was also developed in their study to evaluate sight distance deficiency of various alternative alignments being evaluated for optimization.

A hierarchical approach to alignment optimization is developed in Eungcheol Kim's Ph.D. dissertation (2001) and in Kim *et al.* (2005). This approach first optimizes the alignment roughly for relatively long road sections and then focuses on optimizing it more precisely for each subsection of that alignment.

3.3.5 Summary

It is clear that alignment optimization has attracted considerable interest from many researchers. Numerous aspects of the problem, objectives, constraints, and solution methods have been explored in considerable depth. However, there are still vast opportunities for further research and development on alignment evaluation and optimization. Specific suggestions for areas where further research would be valuable are suggested in Chapter 15.

Part II
Highway Alignment Optimization with Genetic Algorithms

PART II further discusses the genetic algorithms (GAs)-based highway alignment optimization (HAO) model developed by the University of Maryland research team with its core concept, structure, mathematical formulation, and effective constraint handling techniques to improve its computation efficiency. Chapter 4 demonstrates how horizontal and vertical alignments of a highway are represented with GAs. Methods to generate and evaluate highway alignments with custom GA operators are also discussed in this chapter. Chapter 5 discusses HAO model formulation; a basic model structure and various cost functions associated with the model objective function and constraints are demonstrated. Methods for handling infeasible solutions of evolutional algorithms like GAs are discussed in Chapter 6. Finally, effective constraint handling methods customized for the GAs-based HAO model are discussed in Chapters 7 and 8. All notations used in PART II can be found in Appendix A.

Chapter 4

Modeling Highway Alignments with GAs

4.1 Representation of Highway Alignments with GAs

A horizontal alignment of a highway is typically defined by a series of tangents, circular curves, and the connecting transition curve sections. A vertical alignment is defined by the graded-tangents connected with parabolic curves. The configuration of these elements depends on the points of intersections (PI's); thus generating a highway alignment can be reduced to determining its corresponding series of PI's. To find highway alignments that connect points A and B, genetic algorithms (GAs) are useful. In the GA-based highway alignment optimization (HAO) model developed by the authors, the alignments are represented with chromosomes, and each chromosome has a series of genes defined by the xyz coordinates of PI's (see Figure 4.1);

$$\mathbf{PI}_i = (x_i, y_i, z_i) \quad \text{for all } i = 1, \ldots, n_{PI}.$$

It is important to note that the genes are not independent of each other because if a coordinate of one PI is changed, the alignment configuration at other PI's may change. It should also be noted that the alignment optimization process is based on a pool-based search of the GA rather than single solution comparison, and thus a set of possible highway alignments (i.e., chromosomes) is treated as the population in the model. About forty to a hundred alignments are generated each time depending on the complexity of the chromosome (i.e., the number of genes), and the individual alignments within each generation compete with each other to reproduce offspring based on their "fitness" (i.e., objective function value). After a sufficient number of generations, the fittest individuals should survive, whereas poor solutions get discarded, and the population will finally converge to an optimized solution.

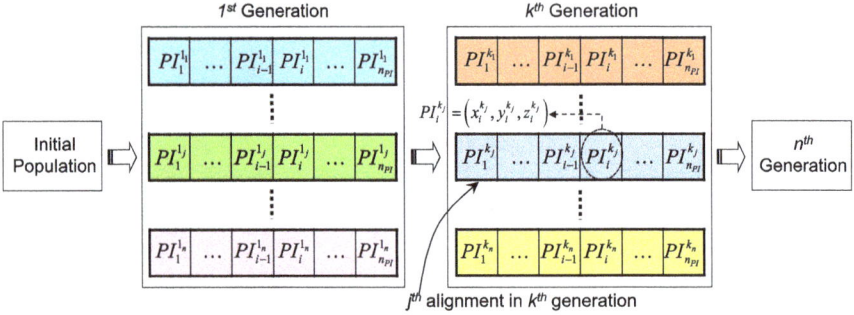

Figure 4.1 Representation of highway alignments in GA

The stopping criterion of the GA optimization method is based on the improvement in the objective function values of the alignments reproduced. Thus, if there is no significant improvement in the objective function value during a certain number of generations (e.g., less than 0.05% improvement during 50 generations), the alignment optimization process is terminated. Eight customized genetic operators have been developed for reproducing highway alignments. The role of these genetic operators is summarized in Table 4.1; further details may be found in Jong and Schonfeld (2003).

Figure 4.2 shows orthogonal cutting planes where the PI's of a highway alignment are generated. The orthogonal cutting planes are equally spaced between the endpoints of the alignment. The total number of the cutting planes (denoted as n_{oc}) and the total number of the PI's (denoted as n_{PI}) are both prespecified as model parameters, and the former can be greater than or equal to the latter. This means that not all the cutting planes are selected for generating PI's. For example, n_{oc} is greater than n_{PI} in Figure 4.2. A series of PI's of the alignment is generated only on the cutting planes selected, and each PI has a unique set of xyz coordinates in its corresponding search space (i.e., selected cutting plane). The selection of the cutting planes is randomly processed at every iteration of alignment generation during the optimization procedure.

In the proposed model, a horizontal PI (HPI) is determined by xy coordinates of a PI whose deflection angle is non-zero (note that no horizontal curve is needed for zero deflection angle at a PI). Thus, the numbers of HPI's that actually produce horizontal curves can be less than or equal to

Table 4.1 Description of customized genetic operators for highway alignment optimization

Operator	Description
Uniform mutation	• Randomly select a PI of a parent alignment (i.e., a gene of a chromosome), and then replace its values with randomly selected numbers to maintain genetic diversity in the population.
Straight mutation	• Randomly select two PI's, and then make the deflection angles of other PI's in between the selected PI's zero in order to straighten the alignment between two randomly selected PI's.
Non-uniform mutation	• Randomly select a PI of a parent alignment, and then replace its values with numbers randomly selected from the adjusted (i.e., reduced) mutation range. This operator is applied at latter generations to refine the alignment.
Whole non-uniform mutation	• This operator applies the non-uniform mutation operator to all the PI's of a randomly selected parent alignment. The resulting offspring will be totally different from its parent.
Simple crossover	• Generate a random integer number k between 1 and n_{PI}, and then cross k^{th} PI of two randomly selected parent alignments.
Two-point crossover	• Exchange the PI's between two randomly generated positions k and l for two randomly selected parent alignments.
Arithmetic crossover	• Randomly select two parent alignments to be crossed, and then generate the offspring as a linear combination of two parent alignments, which guarantees the offspring is always feasible.
Heuristic crossover	• Randomly select two parent alignments and generate a single offspring according to a heuristic rule based on the comparison of their objective function values.

the total number of PI's specified for generating the highway alignments in the model. As with the horizontal curve, there is no vertical curve in the middle PI among three consecutive PI's if they have the same elevation (i.e., z value). There is also no vertical curve at the middle PI if the three PI's are aligned in a sloping straight line. Thus, the number of vertical PI's (VPI's) that actually produce vertical curves can also be less than or equal to the total number of PI's specified for the alignment generation. It is important to note that (1) both a horizontal and a vertical curve, (2) either one of them, or (3) neither of them may be needed at each PI generated. Therefore, a

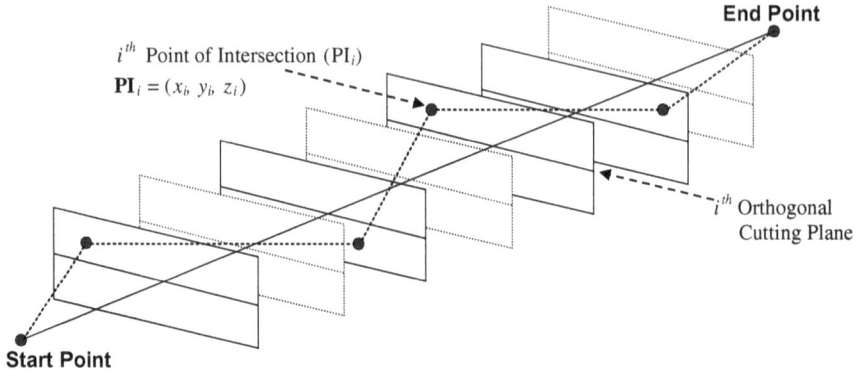

Figure 4.2 A series of PI's generated on corresponding orthogonal cutting planes

subset of the PI's used for generating horizontal curves can differ from that used for the vertical curves.

It is also important to note that the number of PI's in the model is a key input parameter that affects the configuration and total cost of the highway alignments generated. In dense urban areas and areas with significant variation in topography and/or land-use a higher PI density will improve the precision of the alignment generation and cost evaluation, whereas in areas with slight variation in topography or land-use, fewer PI's will suffice. Thus, PI density (i.e., the number of PI's) should be decided on a case-by-case basis in the alignment optimization process with careful consideration of land-use complexity and topography of the project area. An example case study for selecting the preferable number of PI's is presented in Section 12.3.1.

4.2   Modeling Horizontal Alignments

In a highway alignment, a series of tangents and curved sections are adjoined. Circular curves and transition curves are typically combined to form the horizontal curved sections. Some kind of transition curve is often applied between a tangent and a circular curve for mitigating a sudden change in degree of curvature and hence, in lateral acceleration and force, from the tangent to the circular path. Particularly for high-speed highway alignments, spiral transition curves are strongly recommended in horizontal curved sections. Spiral transition curves, which are widely used in practice, are chosen to model the transition curves. For most design standards, the model allows users to easily specify their own input values, but the default values are taken from the relevant AASHTO design manual (AASHTO, 2011).

Figure 4.3(a) presents an example of a typical horizontal alignment with a series of reference points, representing intersection points between tangents, circular curves, and transition curves. For notational convenience, we let $\mathbf{EP}_1$ and $\mathbf{EP}_2$ be the start and end points, respectively, of a highway alignment, and its initial and final PI's (i.e., $\mathbf{PI}_0$ and $\mathbf{PI}_{n_{PI}+1}$, respectively) correspond to the start and end points.

As shown in the figure, $\mathbf{ST}_i$ and $\mathbf{TS}_{i+1}$ are linked by a straight line connecting $\mathbf{PI}_i$ and $\mathbf{PI}_{i+1}$ for all $i = 0, \ldots, n_{PI}$ whereas $\mathbf{TS}_i$ and $\mathbf{SC}_i$ and $\mathbf{CS}_i$ and $\mathbf{ST}_i$ are connected by a spiral transition curve and $\mathbf{SC}_i$ and $\mathbf{CS}_i$ are connected by a circular curve for all $i = 1, \ldots, n_{PI}$. Note that in an extreme case, where an alignment tangent section between two consecutive intersection points [e.g., between $\mathbf{PI}_2$ and $\mathbf{PI}_3$ in Figure 4.3(a)] is completely eliminated by two spiral transition curves, the point of change from spiral to tangent section pertaining to one intersection point will coincide with the point of change from tangent to spiral curve pertaining to the next intersection point [e.g., $\mathbf{ST}_2$ and $\mathbf{TS}_3$ are the same point in Figure 4.3(a)]. Furthermore, if an intersection angle at $\mathbf{PI}_i$ (denoted as θ_{PI_i}) becomes zero, all the reference points pertaining to $\mathbf{PI}_i$ are the same; for example, the locations of $\mathbf{TS}_4$, $\mathbf{SC}_4$, $\mathbf{CS}_4$, $\mathbf{ST}_4$, and $\mathbf{PI}_4$ shown in Figure 4.3(a) would then all be the same.

Figure 4.3(b) shows a typical horizontal curved section of the alignment with two spiral transition curves connecting the central circular curve to the adjoining tangents. Without violating the AAHTO design standards (AASHTO, 2011) the minimum radius can be used to fit the radius of a circular curve (R_{H_i}), and minimum superelevation run-off length can be used to fit the length of spiral transition curve (l_{ST_i}). We now realistically represent the centerline of the horizontal alignment given the coordinates of $\mathbf{PI}_i$ (for $i = 1, \ldots, n_{PI}$) and specified design codes. The coordinates of $\mathbf{TS}_i$, $\mathbf{SC}_i$, $\mathbf{CS}_i$, and $\mathbf{ST}_i$ can be found with simple vector operations shown in Section 4.2.1, Horizontal Alignment Generation Procedure. All notations used in the procedure are summarized in Table 4.2.

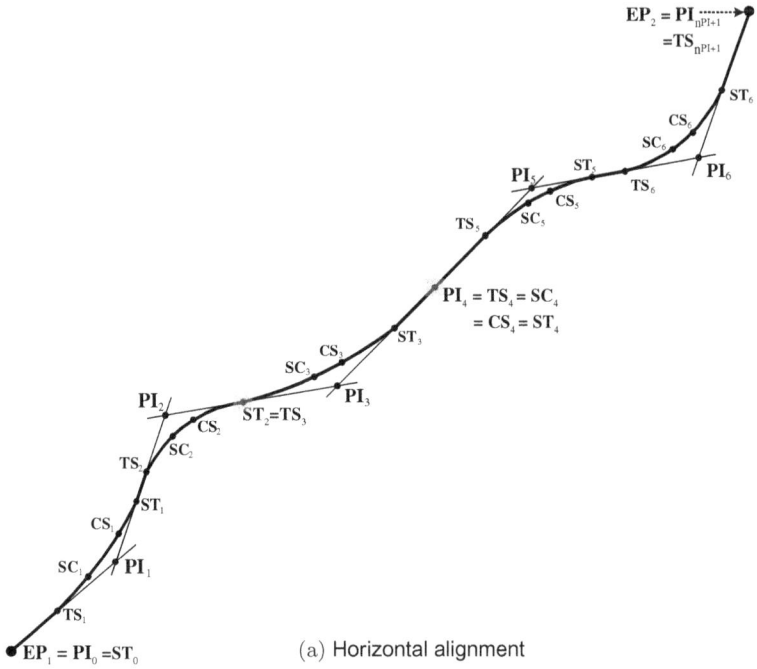

(a) Horizontal alignment

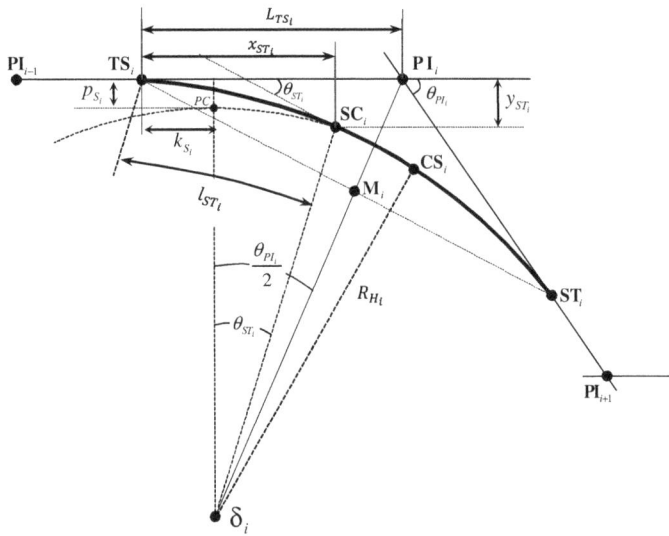

(b) Horizontal curved section

Figure 4.3   Geometric specification of typical horizontal alignments used in the HAO model

Table 4.2　Notation used in horizontal alignment generation procedure

Notation	Descriptions
$\theta_{PI_i} =$	Deflection angle at the ith point of intersection (PI)
$\theta_{ST_i} =$	Spiral angle at the ith horizontal curved section
$\delta_i =$	The center point of the ith horizontal curved section
$CS_i =$	The point of change from circle to spiral pertaining to PI_i
$EP =$	Start or end point of a highway alignment; $EP_1 = PI_0$; $EP_2 = PI_{n_{PI}+1}$
$h_m =$	Minimum vertical clearance
$k_{Si} =$	Abscissa of the shifted PC referred to TS_i
$L_{TS_i} =$	Tangent distance from TS_i to PI_i
$l_{ST_i} =$	The length of spiral transition curve at i^{th} horizontal curve section
$M_i =$	The middle point of the line segment connecting TS_i to ST_i
$n_{OC} =$	Total number of orthogonal cutting planes where PI's are generated
$n_{PI} =$	Total number of PI's that outlines the highway alignment generated
$p_{S_i} =$	Offset from the initial tangent to the PC of the shifted circle
$PI_i =$	i^{th} PI of the highway alignment generated; $PI_i = (x_i, y_i, z_i)$ for $i = 1, \ldots, n_{PI}$
$R_{H_i} =$	Horizontal curve radius at the ith horizontal curve section
$RP =$	Reference points required to model highway structures at endpoints
$R(\theta) =$	Rotation matrix
$SC_i =$	The point of change from spiral to circle pertaining to PI_i
$ST_i =$	The point of change from spiral to tangent pertaining to PI_i
$TS_i =$	The point of change from tangent to spiral pertaining to PI_i
$x_{ST_i} =$	Total tangent distance from TS_i to SC_i with reference to initial tangent
$y_{ST_i} =$	Total tangent offset at SC_i with reference to TS_i and initial tangent

4.2.1　Horizontal Alignment Generation Procedure

STEP 1: Generate PI_i of a new alignment along the orthogonal cutting planes using GA operators, and then connect PI_i and PI_{i+1} with a straight line for $i = 1, \ldots, n_{PI}$

STEP 2: Find deflection angle (θ_{PI_i}) at PI_i for $i = 1, \ldots, n_{PI}$

$$\theta_{PI_i} = \cos^{-1}\left[\frac{(PI_i - PI_{i-1}) \cdot (PI_{i+1} - PI_i)}{\|PI_i - PI_{i-1}\| \, \|PI_{i+1} - PI_i\|}\right] \tag{4.1}$$

where: $\| \, \| =$ length of a vector;

$\cdot =$ dot (inner) product

STEP 3: Find $\mathbf{TS}_i$ and $\mathbf{ST}_i$ for $i = 1, \ldots, n_{PI}$

$$\mathbf{TS}_i = \mathbf{PI}_i + L_{TS_i} \times \frac{(\mathbf{PI}_{i-1} - \mathbf{PI}_i)}{\|\mathbf{PI}_{i-1} - \mathbf{PI}_i\|} \tag{4.2}$$

$$\mathbf{ST}_i = \mathbf{PI}_i + L_{TS_i} \times \frac{(\mathbf{PI}_{i+1} - \mathbf{PI}_i)}{\|\mathbf{PI}_{i+1} - \mathbf{PI}_i\|} \tag{4.3}$$

where: L_{TS_i} = tangent distance from $\mathbf{TS}_i$ to $\mathbf{PI}_i$

$$L_{TS_i} = k_{S_i} + (R_{H_i} + p_{S_i}) \times \tan(\theta_{PI_i}/2) \tag{4.4}$$

STEP 4: Find $\mathbf{M}_i$ and δ_i for $i = 1, \ldots, n_{PI}$

$$\mathbf{M}_i = [\mathbf{TS}_i + \mathbf{ST}_i]/2 \tag{4.5}$$

$$\delta_i = \mathbf{PI}_i + [(R_{H_i} + p_{S_i}) \times \sec(\theta_{PI_i}/2)] \times \frac{(\mathbf{M}_i - \mathbf{PI}_i)}{\|\mathbf{M}_i - \mathbf{PI}_i\|} \tag{4.6}$$

STEP 5: Find $\mathbf{SC}_i$ and $\mathbf{CS}_i$ for $i = 1, \ldots, n_{PI}$

$$\mathbf{SC}_i = R_{H_i} \times \frac{(\mathbf{M}_i - \delta_i)}{\|\mathbf{M}_i - \delta_i\|} \times \mathbf{R}\left(\frac{\theta_{PI_i}}{2} - \theta_{ST_i}\right) \tag{4.7}$$

$$\mathbf{CS}_i = R_{H_i} \times \frac{(\mathbf{M}_i - \delta_i)}{\|\mathbf{M}_i - \delta_i\|} \times \mathbf{R}\left(-\frac{\theta_{PI_i}}{2} + \theta_{ST_i}\right) \tag{4.8}$$

where: $\mathbf{R}(\theta)$ = rotation matrix; $\mathbf{R}(\theta) = \begin{bmatrix} \cos(\theta) & -\sin(\theta) \\ \sin(\theta) & \cos(\theta) \end{bmatrix}$

$\theta_{ST_i} = S_{T_i}/(2R_{H_i})$

STEP 6: Connect $\mathbf{TS}_i$, $\mathbf{SC}_i$, $\mathbf{CS}_i$, and $\mathbf{ST}_i$, (for $i = 1, \ldots, n_{PI}$) with corresponding spiral transition and circular curves.

4.3   Modeling Vertical Alignments

In the HAO model, the horizontal and vertical alignments of a new highway are generated and evaluated jointly; the vertical alignment is determined by fitting parabolic curves to graded-tangents at vertical points of intersections (VPI's) while its corresponding horizontal alignment is being created. The evaluation of the highway (i.e., its total cost and environmental impact calculation) is then processed after its horizontal and vertical alignments are generated.

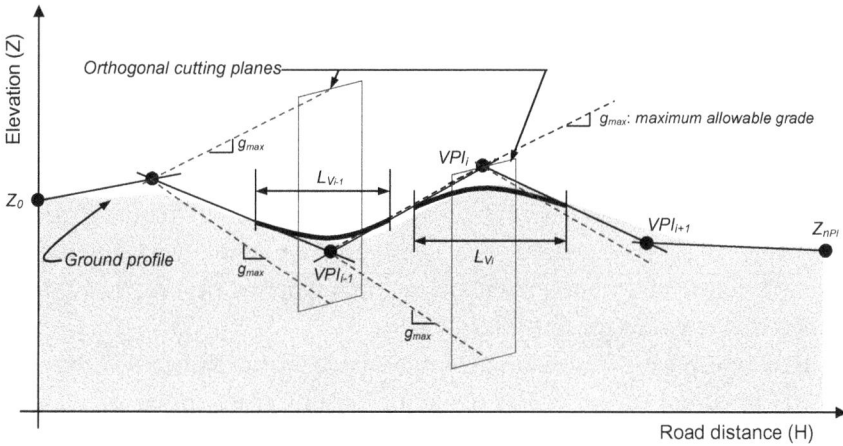

Figure 4.4 Geometric specification of a typical vertical curve section used in the HAO model

Let the *HZ* plane be a coordinate system designed to represent ground and road elevation along the horizontal alignment. The *H* and *Z* axes represent road distance and elevation along the horizontal alignment, respectively (see Figure 4.4). We now define a vertical alignment on the *HZ* plane. Let Z_0 and $Z_{n_{PI}+1}$ be the elevation of start and end points of the vertical alignment, respectively. A series of VPI's, generated from the GA, then can be placed in the *HZ* plane, while considering design constraints. Here, VPI_i can be defined with a pair of H_i and Z_i for all $i = 1, \ldots, n_{PI}$. A series of VPI's are located as close as on a ground profile, and outlines the track of the vertical alignment. Linking each pair of successive points with a straight line produces a piecewise linear trajectory of the vertical alignment. Taking into account all available design constraints (e.g., sight distance, vertical clearance, maximum gradient, and minimum vertical curve length constraints), vertical alignments are generated on the HZ plane (see Figure 4.4).

Note that the horizontal alignment is generated through the Horizontal Alignment Generation Procedure shown in Section 4.2.1 with the set of PI's. The vertical alignment is also created (jointly) with the PI's, while their elevations (i.e., Z values of PI's) are obtained from the Road Elevation Determination Procedure described in Section 7.3.1. More detailed discussion of the vertical alignment generation may be found in Chapter 7.

4.4   Modeling Highway Endpoints

In the earlier version of the HAO model (e.g., HAO 1.0 to HAO 6.0 as described in Table 3.1, Chapter 3), the start and end points of a new highway were assumed to be predetermined by model users before the optimization process. However, such an assumption was relaxed in the most recent version of the HAO model (i.e., HAO 7.0). In this model version the highway endpoints are also considered as decision variables rather than given inputs, and additional information for existing roads where a new highway starts and ends is required for this relaxation.

Here, we make a reasonable assumption that the model users can specify several preferred sub-segments along the existing roads. Such an assumption is realistic since there may be many critical points along the existing roads which are not suitable as junction points between the new and existing roads. For instance, near existing interchanges (or intersections), sharply curved sections in which drivers' sight-distances are insufficient, and bridge sections on existing roads may not be appropriate for the junction points. These critical points should be prescreened before the optimization process. Given the basic information of the existing road identified from a Geographic Information System (GIS) database (e.g., a horizontal profile of the existing road and its corresponding elevation data) and user preferences (i.e., preferred road segments) for the endpoints of a new highway, a method for determining the highway endpoints is described in the next sections.

4.4.1   Determination of Highway Endpoints

It is assumed that a piecewise linear data format is used to save and extract the coordinates of an existing road. This makes it easy to represent the existing road in the model with a simple vector operation. A sufficient number of points may be required for representing the road realistically. Suppose that ten intermediate points are successively specified by the model users along the existing road to which a new highway will be connected (see Figure 4.5(a)). Then, piecewise linear segments obtained from the connection of the intermediate points roughly outline the existing road. The XY coordinates of all the intermediate points can be easily obtained from an

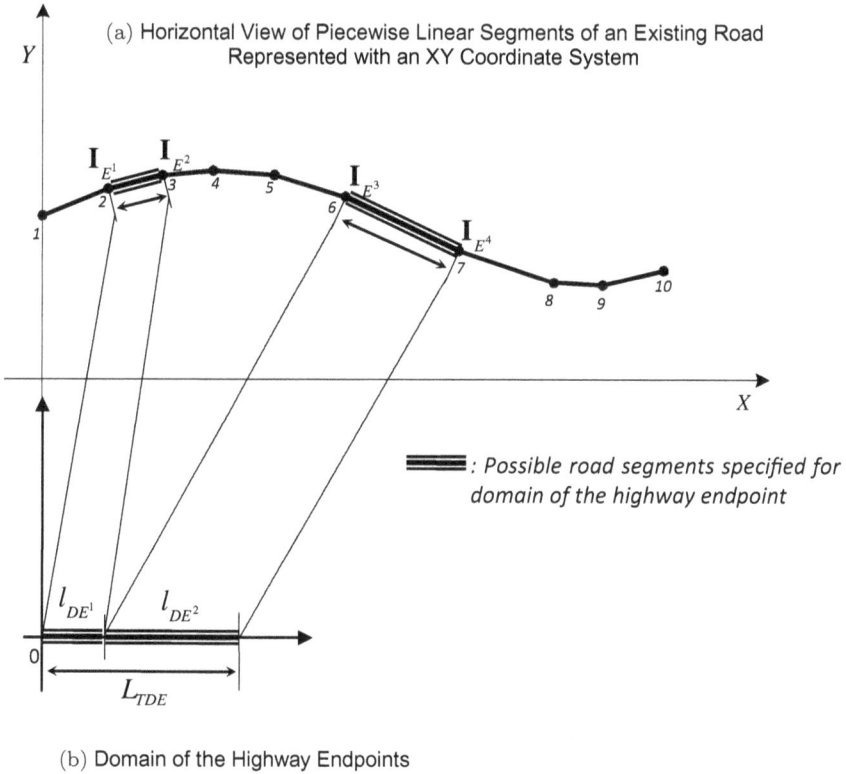

(a) Horizontal View of Piecewise Linear Segments of an Existing Road Represented with an XY Coordinate System

: Possible road segments specified for domain of the highway endpoint

(b) Domain of the Highway Endpoints

Figure 4.5 Representation of domain of the possible endpoints

input GIS database for the study area, and ground elevations of those points may also be directly obtained through a DEM.[1]

It should be noted here that we do not need XY coordinates for all the intermediate points (here ten) to generate the highway endpoints. If two sub-segments of the existing highway are selected for domains of the possible highway endpoint (or start point), as shown in Figure 4.5, only coordinates of the four intermediate points (here I_{E^1}, I_{E^2}, I_{E^3}, and I_{E^4}) must be identified.

---

[1] The digital elevation model (DEM) is the most common basis used in many GISs as a type of digital terrain model (DTM), recording a topographical representation of the terrain of the Earth or another surface in digital format. The DEM normally divides the area into rectangular pixels and stores the elevation of each pixel.

Possible locations of the end point are continuous along the specified road segments and will be generated with the endpoint determination procedure in the following.

Let the length of each road segment specified for an endpoint generation be l_{DE}. Then the domain of the possible endpoints can reduce to $L_{TDE} = \sum_{j=1}^{n_{seg}} l_{DE^j}$, where n_{seg} is total number of road segments specified. Given the ground elevation database (DEM) and XY coordinates of all the specified intermediate points (denoted as $\mathbf{I}_{E^i} = (x_{I_{E^i}}, y_{I_{E^i}})$ for all (i), an algorithm for determining possible locations of the highway endpoint is developed. Note that three additional reference points, which are necessary for representing the three-leg structures of the endpoint, are also found from the following procedure. All notations used in the procedure are summarized in Table 4.3.

Table 4.3 Notation used in endpoint determination procedure

Notation	Descriptions
$\theta_{EP} =$	Intersection angle at the endpoint of the new highway
$\mathbf{I}_{E^i}, \mathbf{I}_{E^{i+1}} =$	A pair of intermediate points specified for representing a preferred road segment of the highway endpoint along the existing road, for $i = 1, \ldots, n_i$
$n_i =$	Total number of intermediate points specified
$l_{DE^j} =$	Length of a road segment, specified for an endpoint generation, for $j = 1, \ldots, n_{seg}$
$n_{seg} =$	Total number of the road segments specified; $n_{seg} = n_i / 2$
$L_{TDE} =$	Total length of the road segments specified
$r_c[A, B] =$	A random value from a continuous uniform distribution whose domain is within the interval $[A, B]$
$l_{Temp} =$	A provisional random value from $r_c[A, B]$
$k =$	The road segment selected for the highway endpoint from the random search process
$\mathbf{I}_{E^{2k-1}}, \mathbf{I}_{E^{2k}} =$	A pair of intermediate points corresponding to the selected road segment k for the highway endpoint
$l_{seg} =$	Distance from $\mathbf{I}_{E^{2k-1}}$ to the highway endpoint
$\Delta l_{seg} =$	A provisional distance used for finding the reference points $\mathbf{RP}_1$ and $\mathbf{RP}_2$; (typically less than 10 ft)
$h_m =$	Minimum vertical clearance for grade separation
$\mathbf{EP} =$	A 3D point found for the highway endpoint (either start or end point); $\mathbf{EP} = (x_{EP}, y_{EP}, z_{EP})$
$\mathbf{RP} =$	Reference points required to model highway structures at endpoints; $\mathbf{RP}_0 = (x_{RP_0}, y_{RP_0}, z_{RP_0})$, $\mathbf{RP}_1 = (x_{RP_1}, y_{RP_1}, z_{RP_1})$, $\mathbf{RP}_2 = (x_{RP_2}, y_{RP_2}, z_{RP_2})$

4.4.2 Highway Endpoint Determination Procedure

STEP 1: For all pairs of the intermediate points specified, calculate l_{DE^j} and L_{TDE}

For $i = n_i - 1$

$$l_{DE^j} = \|\mathbf{I}_{E^{i+1}} - \mathbf{I}_{E^i}\| = \sqrt{(x_{I_{E^{i+1}}} - x_{I_{E^i}})^2 + (y_{I_{E^{i+1}}} - y_{I_{E^i}})^2} \quad (4.9)$$

$i = i + 1$

End

$$L_{TDE} = \sum_{j=1}^{n_{seg}} l_{DE^j} \quad (4.10)$$

STEP 2: Find l_{Temp} randomly between 0 and L_{TDE}

$$l_{Temp} = r_c[0, L_{TDE}] \quad (4.11)$$

STEP 3: Find selected segment k and l_{seg}

if $l_{Temp} \le l_{DE^1}, \rightarrow l_{seg} = l_{Temp}, k = 1$ and go to STEP 4
else
    for $i = 2$ to n_{seg}
        if $l_{Temp} \le \sum_{j=1}^{i} l_{DE^j} \rightarrow l_{seg} = l_{Temp} - \sum_{j=1}^{i-1} l_{DE^j}, k=1$ and go to STEP 4
        else $\rightarrow i = i + 1$
    end
end

STEP 4: Find XY coordinates of three reference points $(\mathbf{RP}_0, \mathbf{RP}_1, \mathbf{RP}_2)$ required for modeling structures of the highway endpoints

$$\begin{bmatrix} x_{RP_0} \\ y_{RP_0} \end{bmatrix} = \mathbf{I}_{E^{2k-1}} + l_{seg} \times \frac{\mathbf{I}_{E^{2k}} - \mathbf{I}_{E^{2k-1}}}{\|\mathbf{I}_{E^{2k}} - \mathbf{I}_{E^{2k-1}}\|} \quad (4.12)$$

$$\begin{bmatrix} x_{RP_1} \\ y_{RP_1} \end{bmatrix} = \mathbf{I}_{E^{2k-1}} + (l_{seg} - \Delta l_{seg}) \times \frac{\mathbf{I}_{E^{2k}} - \mathbf{I}_{E^{2k-1}}}{\|\mathbf{I}_{E^{2k}} - \mathbf{I}_{E^{2k-1}}\|} \quad (4.13)$$

$$\begin{bmatrix} x_{RP_1} \\ y_{RP_1} \end{bmatrix} = \mathbf{I}_{E^{2k-1}} + (l_{seg} + \Delta l_{seg}) \times \frac{\mathbf{I}_{E^{2k}} - \mathbf{I}_{E^{2k-1}}}{\|\mathbf{I}_{E^{2k}} - \mathbf{I}_{E^{2k-1}}\|} \quad (4.14)$$

STEP 5: Find Z coordinates of the three reference points

$\rightarrow$ Compute ground elevations of the three reference points (z_{RP_0}, z_{RP_1}, z_{RP_2}), using a planar-interpolation method given with the GIS elevation data base (e.g., DEM)

STEP 6: Find a 3D highway endpoint (either start or end point)

$\rightarrow$ $\mathbf{EP} = \mathbf{RP}_0 = \begin{bmatrix} x_{RP_0} \\ y_{RP_0} \\ z_{RP_0} \end{bmatrix}$, Case 1: for at-grade intersected structures

$\rightarrow$ $\mathbf{EP} = \mathbf{RP}_0 = \begin{bmatrix} x_{RP_0} \\ y_{RP_0} \\ z_{RP_0} + h_m \end{bmatrix}$, Case 2: for grade-separated structures with overpassing roads

$\rightarrow$ $\mathbf{EP} = \mathbf{RP}_0 = \begin{bmatrix} x_{RP_0} \\ y_{RP_0} \\ z_{RP_0} - h_m \end{bmatrix}$, Case 3: for grade-separated structures with under-passing roads

In Step 1, the length of each road segment defined by the model users is calculated, and then a temporary random value is generated from the uniform random process with domain 0 to L_{TDE} through Step 2. From Step 3 to Step 4, XY coordinates of the three reference points are computed with a simple vector operation. In Step 5, elevations (Z coordinates) of the reference points are obtained through a planar interpolation method, given XY coordinates of those points and the input GIS elevation database (DEM). Note that we may obtain the elevations directly from the provided input DEM without the interpolation method; however, this is less desirable since they may not be sufficiently accurate to use directly in the alignment design process.

Finally, XYZ coordinates of the highway endpoint EP (either start or end point of the new alignment) are computed based on those of the reference point $\mathbf{RP}_0$ found in Steps 4 and 5. Elevation (Z_{EP}) of $\mathbf{EP}$ can be varied depending on type of structures considered for representing the endpoint. If an at-grade intersection (e.g., three-leg intersection or roundabout) is considered, the XYZ coordinates of $\mathbf{EP}$ become those of $\mathbf{RP}_0$. However, if a grade separated structure (e.g., a trumpet interchange) is considered, the XY coordinates of $\mathbf{EP}$ become those of $\mathbf{RP}_0$, while its ground elevation is either elevated to $Z_{RP_0} + h_m$ (for over-passing the existing road) or lowered to

$z_{RP_0} - h_m$ (for under-passing the existing road). Note that type of structures used for representing the endpoint and their minimum vertical clearance (h_m) can be specified by the model users.

4.4.3 GA Operators for Endpoint Generation

Three customized GA operators are employed for evolving the highway endpoints during the alignment search process. These are:

- Uniform mutation
- One-point crossover
- Two-point crossover

The uniform mutation operator is used for arbitrarily altering the endpoints (either start or end points) of selected chromosomes.[2] Let the chromosome to be mutated be $\Lambda = [\mathbf{EP}_1, \mathbf{PI}_1, \ldots, \mathbf{PI}_n, \mathbf{EP}_2]$ and either $\mathbf{EP}_1$ or $\mathbf{EP}_2$ be selected to apply the uniform mutation. Then, the EP will be replaced with a new endpoint from the random search process of the Highway Endpoint Determination Procedure Figure 4.6(a) shows a good example when this mutation operator works well during the optimization process. As shown in the figure, a good offspring is generated by replacing the start point ($\mathbf{EP}_1^i$) of a selected parent alignment with new one ($\mathbf{EP}_1^{i+1}$), while inheriting the other genes (i.e., a set of PI's and $\mathbf{EP}_2^i$) from the parent. The resulting offspring completely avoids a no-go area (e.g., environmentally sensitive area) in the search space after the endpoint mutation.

Besides the mutation operators, two crossover operators (one-point and two-point) are used for endpoints evolution. These operators allow the new offspring to inherit good genes from the parents by swapping their endpoints. The concept of the one-point crossover is to exchange only one endpoint (either the start or endpoint) between two selected parents, while two-point crossover swaps both endpoints of the two parents simultaneously. Suppose that two parents (Λ^1 and Λ^2) are selected for the crossover operation, where $\Lambda^1 = [\mathbf{EP}_1^1, \mathbf{PI}_1^1, \ldots, \mathbf{PI}_n^1, \mathbf{EP}_2^1]$ and $\Lambda^2 = [\mathbf{EP}_1^2, \mathbf{PI}_1^2, \ldots, \mathbf{PI}_n^2, \mathbf{EP}_2^2]$. Then, the resulting offspring from the one-point

---

[2]The highway alignments are represented with chromosomes in the GA-based HAO model (see Section 4.1). Readers may refer to Jong (1998) and Jong and Schonfeld (2003) for a detailed description of the GA encoding for highway alignments.

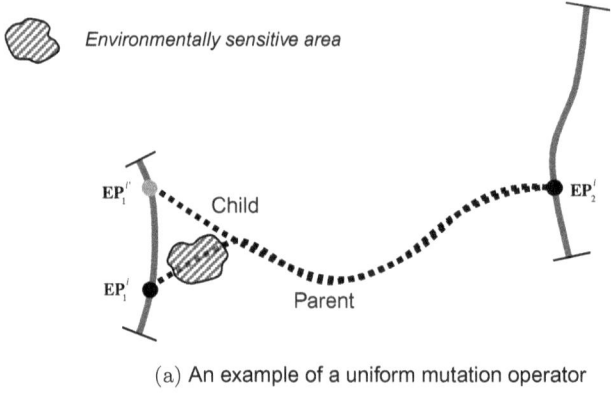

(a) An example of a uniform mutation operator

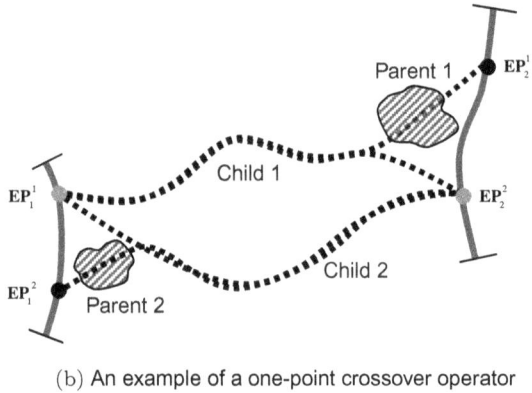

(b) An example of a one-point crossover operator

Figure 4.6   Examples of endpoint operators

crossover operator are:

$$\Lambda^{1'} = [\mathbf{EP}_1^1, \mathbf{PI}_1^1, \ldots, \mathbf{PI}_n^1, \mathbf{EP}_2^2] \quad \Lambda^{2'} = [\mathbf{EP}_1^1, \mathbf{PI}_1^2, \ldots, \mathbf{PI}_n^2, \mathbf{EP}_2^2]$$

$$\text{or} \qquad\qquad\qquad \text{or}$$

$$= [\mathbf{EP}_1^2, \mathbf{PI}_1^1, \ldots, \mathbf{PI}_n^1, \mathbf{EP}_2^1] \quad [\mathbf{EP}_1^2, \mathbf{PI}_1^2, \ldots, \mathbf{PI}_n^2, \mathbf{EP}_2^1]$$

The resulting offspring from two-point crossover are:

$$\Lambda^{1'} = [\mathbf{EP}_1^2, \mathbf{PI}_1^1, \ldots, \mathbf{PI}_n^1, \mathbf{EP}_2^2]$$

$$\Lambda^{2'} = [\mathbf{EP}_1^1, \mathbf{PI}_1^2, \ldots, \mathbf{PI}_n^2, \mathbf{EP}_2^1]$$

Figure 4.6(b) shows a successful example of the one-point crossover operator. Child 1 ($\Lambda^{1'}$) inherits its endpoint from parent 2 (i.e., $\mathbf{EP}_2^2$), while the other genes are inherited from parent 1 (Λ^1). Similarly, child 2 ($\Lambda^{2'}$) inherits its start point from parent 1 (i.e., $\mathbf{EP}_1^1$), while the other genes are taken over from parent 2 (Λ^2). More detailed genetic encoding of the GA operators employed for the endpoint generation may be found in Jong (1998).

4.5 Modeling Highway Structures

Many highway structures may also be included in constructing new highways. These may include bridges for crossing rivers or valleys, tunnels, and cross-structures for intersecting existing highways (e.g., at-grade intersections, grade separation, and interchanges). Especially in mountainous terrain and river valleys, construction of tunnels and bridges may dominate the highway design process. Hence, such structures should also be considered in the highway alignment optimization process.

The basic principle for locating highway bridges or tunnels (unless they are major) is that the highway location should normally determine the structure location, not the reverse. If the bridge or tunnel location is located first, in most cases the resulting highway alignment is not the best. Thus, the general procedure for the highway design should be to first determine the highway location and next determine the bridge and/or tunnel sites or consider both simultaneously (Garber and Hoel, 1998). In the HAO model, we consider circumstances where bridge or tunnel construction is more economical than earthwork. During the alignment optimization process, an economic break-even point between earthwork cost and construction cost for bridges or tunnels is determined based on the elevation difference between the ground and the centerline of the highway alignment, and used for evaluating the alignments generated. The characteristics affecting a small tunnel cost may include tunnel length, cross section, clearance, horizontal alignment, and grade. Factors affecting the cost of bridges for crossing rivers or valleys may include the bridge span length and pier height. A methodology that selects bridge or tunnel constructions in lieu of earthwork (if they are more economical) may be found in the authors' previous publication (Kim *et al.*, 2007), and simple tunnel and bridge cost functions used in the model are presented in Section 5.4, Highway Cost Formulation.

Small highway bridges for grade separation and structures for highway junctions (e.g., interchanges and at-grade intersections) are also considered in the model to evaluate cases where the generated alignments cross existing roads. To model those structures, various design codes associated with the existing and new highways (e.g., minimum vertical clearances, widths of two cross roads, design speeds and design vehicles for turn roads), fill and/or cut slopes, intersection angle between the cross roads, horizontal and vertical PI's of the new alignment adjacent the crossing point, and elevation data of the crossing point are used. Modeling of these structures is further discussed.

Retaining walls (which resist the lateral pressure of soil at highway cut sections) and noise barriers (which block traffic noise from the highway) should also be considered in optimizing highway alignments because locations where those structures are needed and their construction cost may vary significantly with a slight change of highway alignment geometry. Automated methods for determining where such structures are desirable but not yet incorporated in the HAO model.

4.5.1 Small Highway Bridges for Grade Separation

A small highway bridge structure is used for grade separation where two highways cross. Such a structure can be used not only for grade separation, without any connection to the existing roads being crossed, but also for the part of interchange structures connecting the roads. Normally, for small highway bridges, few piers are used to support spans, and they are equally spaced. In addition, the pier heights may be considerably shorter than for bridges crossing rivers. The pier heights of small highway bridges should satisfy the minimum vertical clearance recommended by AASHTO (2011).

Figure 4.7 shows two small highway bridges modeled in the HAO model for grade separation of new and existing roads: (1) a bridge on the new highway where it overpasses an existing road and (2) one on an existing road where the new highway is under-passed. To model the bridge section, various design codes associated with the existing and new highways (e.g., a minimum vertical clearance, fill or cut slope, width of two cross

(a) New highway overpassing an existing road

(b) New highway under-passing an existing road

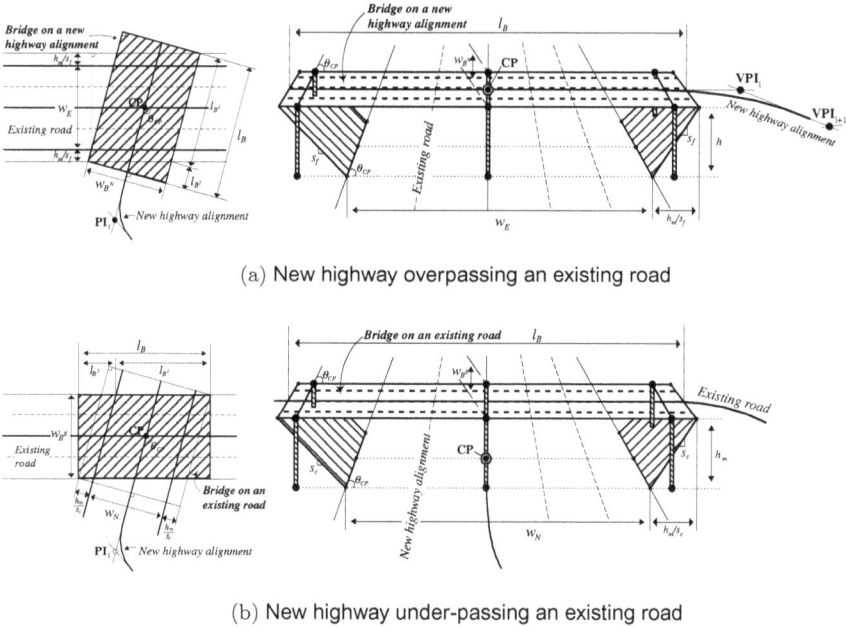

Figure 4.7 Small highway bridges for grade separation of existing roads

roads), intersection angle between the cross roads, horizontal and vertical PI's of the new alignment adjacent to the crossing point are used. Note that a decision on whether the new highway under-passes or overpasses the existing roads is made by comparing the total earthwork cost and structure cost of the entire alignment for each case. The length of the small highway bridge is calculated with Eq. (4.15a) and Eq. (4.15b), and can be used for estimating the bridge structure cost discussed in Section 5.4, Highway Cost Formulation). Notations used in modeling highway structures are presented in Table 4.4.

$$l_B = \left(\frac{w_E + 2h_m/s_f}{\theta_{CP} - \pi/2} \right) + \left(w_{B^N} \tan \left(\theta_{CP} - \frac{\pi}{2} \right) \right),$$

if overpassing an existing road (4.15a)

$$l_B = \left(\frac{w_N + 2h_m/s_c}{\theta_{CP} - \pi/2} \right) + \left(w_{B^E} \tan \left(\theta_{CP} - \frac{\pi}{2} \right) \right),$$

if under − passing an existing road (4.15b)

Table 4.4 Notation used in modeling highway structures

Notation	Descriptions
$\theta_{CP} =$	Intersection angle between two cross roads
$\theta_{EP} =$	Intersection angle at a new highway endpoints
$l_{Bi} =$	Length of the ith highway bridge
$l_{EP} =$	Length of a roadway or ramp associated three-leg structures at the highway endpoint
$s_c, s_f =$	Cut and fill slops, respectively
$w_B =$	Bridge width
$w_E =$	Width of an existing road intersected by the new highway
$w_L =$	Travel lanes width of the new highway
$w_N =$	Width of the new highway; $w_N = w_L + w_S$
$w_P =$	Width of paved portion of the new highway; $w_P \leq w_N$
$w_S =$	Shoulders width of the new highway

4.5.2 Structures for Highway Junction Points with Existing Roads

For optimizing junction points of a new highway with existing roads, the HAO model allows its users to specify several preferred sub-segments along the existing roads. Such an assumption is realistic in the highway design process since there may be some critical points or locations along the existing roads where junctions between new and existing roads are not permitted. For instance, points near interchanges (or intersections), sharply curved sections in which drivers' sight-distances are insufficient, and bridge sections on existing roads may be unsuitable for junctions. These critical points should be prescreened before the optimization process. Given such user preferences, design standards for structures and basic information about the existing road (e.g., a horizontal profile of the existing road and its corresponding elevation data), the HAO model generates simple highway structures that represent junctions of the new highway in addition to its alignments. Note that a piecewise linear data format is used to save and extract the coordinates of existing roads.

Among many types of three-leg cross structures where a new highway diverges from an existing road, a simple trumpet interchange and three-leg at-grade intersection (which are most commonly used in highway design process) are considered in the model. Figure 4.8 shows the centerline of the two three-leg structures employed for representing the endpoint of a new

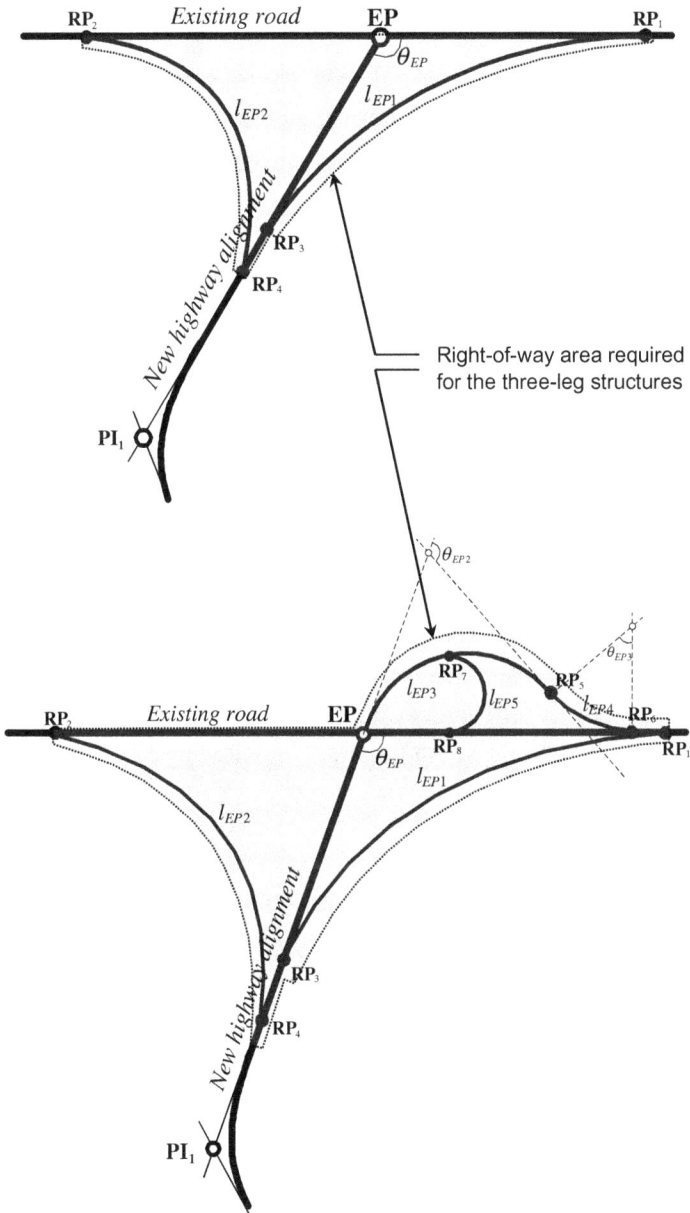

Figure 4.8 Simple three-leg structures represented with a set of reference points

highway on the existing road. Many reference points (at least six and ten for the at-grade intersection and trumpet interchange, respectively) which include the endpoint and the first or last PI of the new highway alignment, are needed to draw the centerlines of the three-leg structures (see Figure 4.8). Such reference points can also be found with simple vector operations as in the highway alignment generation process if design codes for the structures (e.g., minimum vertical clearance, width of the cross roads, design speed and design vehicle for turn roads, fill and/or cut slopes) and the profiles of the existing and new roads are provided.

The at-grade intersection has two separated right-turn ramps; the trumpet interchange comprises a small highway bridge for grade separation and several ramps. Pavement cost, right-of-way cost, and earthwork cost are major construction cost components of the at-grade intersection. For estimating the interchange structure cost, the cost of small highway bridges should be added to the cost items estimated for at-grade intersections. These cost items can be approximately estimated with the horizontal and vertical profiles of roadways associated with the three-leg structures and their cross-section information (see Section 5.4 for more details of cost functions used in the HAO model). Note that the range of the intersection angle (θ_{EP}) at the crossing point is restricted to $\pi/3 \le \theta_{EP} \le 2\pi/3$ based on AASHTO (2011), which prohibits a sharp cross-angle below $\pi/3$.

Simple four-leg cross-structures, such as Diamond and Clover-typed interchanges and at-grade intersections are also considered in the model. Costs of those structures can also be estimated based on roadway profiles associated with the four-leg structures. Other complex and large interchanges are omitted here since they require their own vast research areas.

Chapter 5

Highway Alignment Optimization Formulation

5.1 Objective Function

The HAO model searches for the optimized solution by minimizing the comprehensively formulated objective function, while satisfying design, environmental, and geographic constraints. The objective function is usually a total cost function (C_{Total}) that comprises ten major components: length-dependent cost (C_L), right-of-way cost (C_R), earthwork cost (C_E), structure cost (C_S), maintenance cost (C_M), travel time cost (C_T), vehicle operation cost (C_V), accident cost (C_A), penalty cost (C_P), and environmental cost (C_{EN}). Table 5.1 summarizes the notation used in the HAO model objective function. All these costs are dominating and sensitive to the PI's of highway alignments, and they are formulated as functions of the PI's directly or indirectly in the model (i.e., PI's are decision variables in the highway location selection and alignment optimization problem). A basic formulation for minimizing total cost of highway alignments can be expressed as follows:

$$\underset{x_1, y_1, z_1, \ldots, x_{nPI}, y_{nPI}, z_{nPI}}{Minimize} \quad C_{Total} = (C_L + C_R + C_E + C_S + C_M)$$

$$+ (C_T + C_V + C_A) + C_P + C_{EN} \quad (5.1)$$

Subject to:

$$x_{LBi} \leq x_i \leq x_{UBi}; \quad y_{LBi} \leq y_i \leq y_{UBi}; \quad z_{LBi} \leq z_i \leq z_{UBi}$$
$$\text{for all } i = 1, \ldots, n_{PI} \quad (5.2)$$

$$R_{H_m} < R_{H_i}; \quad S_{SD} < S_{H_i}; \quad S_{T_m} < S_{T_i}$$
$$\text{for all } i = 1, \ldots, n_{HC} \quad (5.3)$$

$$L_{V_m} < L_{V_i}; \quad S_{SD} < S_{V_i}; \quad |g_i| < g_{max_i}$$
$$\text{for all } i = 1, \ldots, n_{VC} \quad (5.4)$$

$$A_k < Max A_k$$
$$\text{for all } k = 1, \ldots, n_{PC} \quad (5.5)$$

where $PI_i = (x_i, y_i, z_i)$ for all $i = 1, \ldots, n_{PI}$

55

Table 5.1 Notation used in the HAO model objective function

Notation	Descriptions
$A_k =$	Affected area of the kth land parcel by the highway alignment generated
$C_A =$	Present value of total accident cost
$C_{E=} =$	Earthwork cost for a new highway alignment development
$C_{EN} =$	Environmental cost for a new highway alignment development
$C_{L=} =$	Length-dependent cost for a new highway alignment development
$C_M =$	Present value of total maintenance cost for a new highway alignment development
$C_P =$	Penalty cost for a new highway alignment development
$C_R =$	Right-of-way cost for a new highway alignment development
$C_S =$	Total structure cost for a new highway alignment development
$C_{Total} =$	Present value of total cost for a new highway alignment development
$C_V =$	Present value of total vehicle operation cost
$C_T =$	Present value of total travel time cost
$C_{T\_Agency} =$	Total agency cost for a new highway development
$C_{T\_User} =$	Total user cost for a new highway development
$g_i =$	Forward or back tangent grade at the ith vertical curve section
$g_{max} =$	Maximum allowable gradient
$L_{V_i} =$	Vertical curve length at the ith vertical curve section
$L_{V_m} =$	Minimum length of vertical curve
$MaxA_k =$	Maximum allowable area of the kth land parcel for the new highway construction
$n_{HC} =$	Total number of horizontal curve sections of the highway alignment generated
$n_{PC} =$	Total number of land parcels affected by the highway alignment generated
$n_{VC} =$	Total number of vertical curve sections in the highway alignment generated
$R_{H_m} =$	Minimum horizontal curve radius
$S_{H_i} =$	Horizontal sight distance at the ith horizontal curve section
$S_{SD} =$	Minimum safe sight distance (i.e., stopping sight distance, given a design speed)
$S_{T_i} =$	Spiral transition curve length at the ith horizontal curve section
$S_{Tm} =$	Minimum length of spiral transition curve
$S_{V_i} =$	Vertical sight distance at the ith vertical curve section
$x_{LB}, y_{LB}, z_{LB} =$	Lower bounds of x, y, z coordinates of PI_i
$x_{UB}, y_{UB}, z_{UB} =$	Upper bounds of x, y, z coordinates of PI_i

x_{LBi}, y_{LBi} *and* z_{LBi} *are lower limits of* i^{th} *PI's* x, y, *and* z *coordinates, respectively.*

x_{UBi}, y_{UBi} *and* z_{UBi} *are upper limits of* i^{th} *PI's* x, y, *and* z *coordinates, respectively.*

R_{H_m}, S_{H_m}, S_{T_m}, L_{V_m}, S_{V_m}, g_{max} can be obtained and calculated with design codes.

L_{Vm} and S_{Vm} may be different for the crest and sag vertical curves.

The first five components (i.e., $C_L + C_R + C_E + C_S + C_M$) of the objective function are mainly incurred by highway agencies, and thus can be classified as the highway agency cost ($C_{T_{Agency}}$). The next three items (i.e., $C_T + C_V + C_A$) can be classified as the user cost ($C_{T_{User}}$), and the second from the last (i.e., C_P) is a penalty cost for alignments that violate constraints. Environmental cost (C_{EN}), which includes various environmental impacts expected along the proposed highway such as air, noise, and water pollution costs, can also be added in the objective function if relevant information (such as traffic volume and unit environmental cost per vehicle mile traveled (VMT)) is available. The definitions and underlying concepts for these cost components are discussed in Section 5.4.

5.2 Constraints

Two types of model constraints are considered in the highway location selection and alignment optimization problem. These are (i) design constraints and (ii) environmental and geographical constraints. The design constraints are used to ensure that the candidate alignments satisfy design standards from AASHTO (2011). The environmental and geographical constraints are used to represent subjective road-project issues (e.g., untouchable and preferred areas for right-of-way boundary of the new highway) that should be satisfied for all highway alternatives. These constraints are sensitive to the topography of the project area, preferences of highway planners and designers, and opinions from public hearings. Thus, the HAO model is designed so that such subjective constraints are provided by the model users (i.e., user specifiable), while the design constraints are governed by the AASHTO standards.

Table 5.2 summarizes the alignment constraints used in the optimization problem. The minimum horizontal curvature constraint, which is the first constraint in the table, depends on design speed, superelevation, and

Table 5.2 Constraints for the highway alignment optimization problem

Category		Type of Constraints
Design constraints	Horizontal alignments	1. Minimum horizontal curvature constraint $(R_{H_m} < R_{H_i})$ 2. Minimum horizontal sight offset at a horizontal curve $(S_{SD} < S_{H_i})$ 3. Minimum superelevation runoff length, only if transition curves are used $(S_{T_m} < S_{T_i})$
	Vertical alignments	4. Minimum vertical curvature constraints $(L_{V_m} < L_{V_i}; S_{SD} < S_{V_i})$ 5. Maximum gradient constraint $(|g_i| < g_{max_i})$ 6. Minimum vertical clearance constraint
Environmental and geographic constraints		7. Environmentally sensitive areas $(A_k < MaxA_k)$

coefficient of side friction of the new highway (see AASHTO, 2011). The second constraint (i.e., the minimum horizontal sight offset) is to make sure that the distance that drivers are able to see at a horizontal curve section is greater than a minimum safety distance (e.g., a stopping sight distance) under a given design speed. The third one, the minimum superelevation runoff length depends on superelevation, maximum relative gradient (in percent), number of lanes rotated, and width of the road. The fourth constraint, the minimum vertical curvature constraint, restricts the length of crest vertical curves to be met with the vertical sight distance as well as guarantees headlight distance and motorist comfort on sag vertical curves. Note that in the model, the solution alignments that violate any of these four constraints, which are associated with either horizontal curved sections or vertical curved sections of the alignments, are processed with the Prescreening and Repairing (P&R) method (see Chapter 8).

The fifth constraint in Table 5.2, i.e., the maximum gradient constraint, normally depends on the nature and importance of the new highway, design speed, and topography of the study area (AASHTO, 2011), and it is specified by the model users. The sixth constraint, minimum vertical clearance is used for restricting road elevation where the new highway intersects existing roads or rivers. For grade separation with an existing road, the minimum required elevation difference between the new and existing roads may be found in AASHTO (2011); however, that required for a bridge crossing a river may vary depending on its water level information (e.g., 100 year floodplains). Note that a Vertical Feasible Gate (VFG) approach, presented in Section 7.3, is designed to represent these gradient and fixed-point constraints in the optimization process.

The environmental and geographical constraints (the seventh in Table 5.2) are too complex to formulate with any single mathematical form since they vary subjectively for various user specifications. When considering roadway construction in a given project area, various geographically sensitive regions (such as historic sites, flood plains, and public facilities) may exist. These control areas should be avoided by the new alignment and to the extent possible, its impact to these regions should be minimized. An effective way of representing such complex constraints in a machine readable format, by developing a Horizontal Feasible Gate (HFG) approach, is presented in Section 7.2.

Additionally, candidate alignments generated from the HAO model also definitely satisfy the alignment boundary conditions, the alignment necessary conditions, the continuity condition, and continuously differentiable conditions, besides the seven distinct constraints shown in Table 5.2. Readers may refer to Jong (1998) for detailed discussion of these additional conditions.

5.3 Integrating GAs and Geographic Information System

In the HAO model, a GA with a number of specialized genetic operators is used for optimizing 3D highway alignments. In addition, a Geographic Information System (GIS) is integrated with the GA to evaluate realistically and comprehensively the highway alignments generated. The primary roles of the GA-based optimization and the GIS module embedded in the model are summarized in Table 5.3.

During the alignment search process, the GA and GIS communicate by exchanging their inputs and outputs (see Figure 5.1). First, a set of new highway alignments (i.e., initial population) is generated from the GA. Then, spatial information about the alignments is transmitted to the GIS, and the alignments' right-of-way cost, environmental impact, and socio-economic impact are evaluated in the GIS, while the other alignment-sensitive costs (e.g., earthwork cost and maintenance cost) are

Table 5.3 Summary of principal processes in the HAO model

Principal process	Role of the principal process
GA-Based Optimization	• Generating highway alignments • Evaluating major alignment-sensitive costs — Earth work, length-dependent, structure cost, and maintenance costs — Travel time, vehicle operation, and accident costs • Searching optimized highway alignments based on the principles of natural evolution and survival of the fittest
GIS-Based Evaluation	• Evaluating alignment's right-of-way cost • Evaluating alignment impacts on the study area — Environmental impact — Socio-economic impact

evaluated in the GA module. After all the costs and impacts of the alignment set are estimated, they are ranked based on their fitness values (i.e., objective function value). Next, the fittest individuals (i.e., alignments ranked with higher fitness values) survive to reproduce new population of the next generation, whereas the least-fit individuals are eliminated. All these evolutionary steps (i.e., alignments generation, evaluation, ranking, and reproduction procedures) repeat until a specified stop-condition is reached.

Three main types of inputs are needed for optimizing the 3D highway alignments: 1) The design specifications, normally defined based on AASHTO (2011) design standards, are needed for generating the highway alignments, which are evaluated based on several unit costs (e.g., unit pavement cost and unit earthwork cost) defined by the model users. 2) The GIS inputs are essential for computing an alignment's right-of-way cost as well as for evaluating its impacts on environmentally and socio-economically sensitive areas in the study region. The model users can also express their preferences, by specifying their areas of interest and untouchable areas in the GIS layers. Finally, 3) information about current and future traffic on the new highway is also needed for user cost estimation.

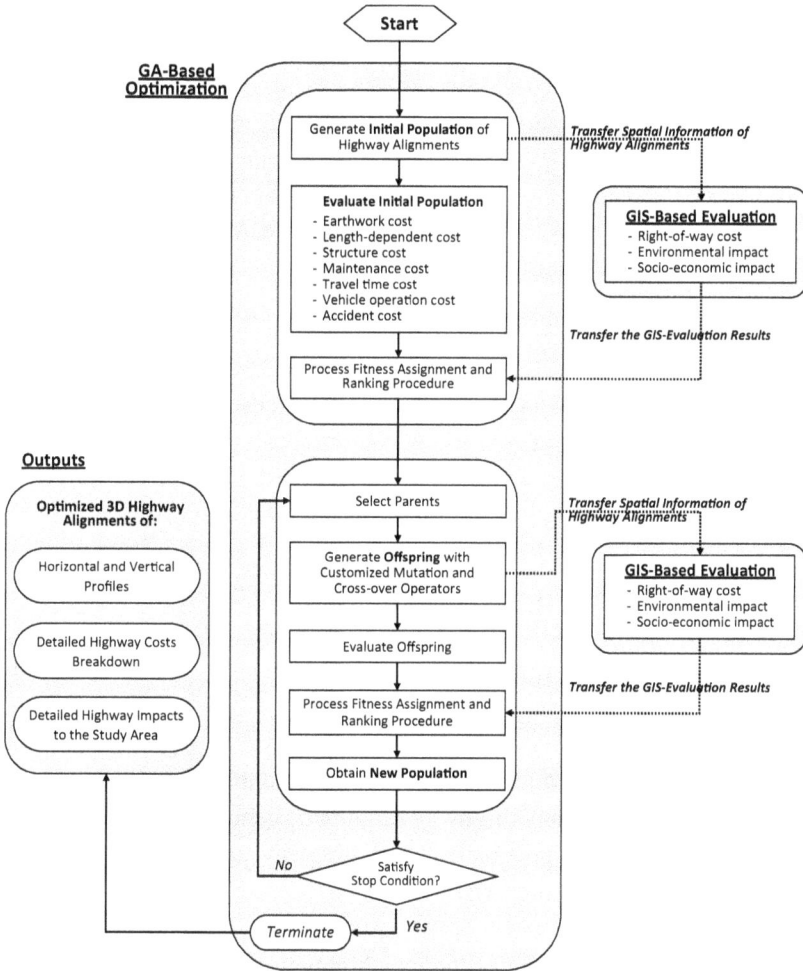

Figure 5.1 GA-GIS based HAO model structure

Various practical and quantitative results of the optimized alignments are provided as model outputs. The model output includes horizontal and vertical profiles of the optimized highway alignments, impacts on the study areas, and various alignment-sensitive costs. These are quite useful to the decision-makers for identifying and refining new highways. A graphical view of the optimized alignments is also provided on a GIS map as a model output.

5.4 Highway Cost Formulation

5.4.1 Highway Agency Cost

As discussed in the previous section, the agency cost associated with a new highway development includes: length-dependent cost (C_L), right-of-way cost (C_R), earthwork cost (C_E), structure cost (C_S), maintenance cost (C_M). Mathematical formulations of the five agency cost components are discussed in the next subsections.

5.4.1.1 *Length-dependent cost* (C_L)

The length-dependent cost (C_L) is defined as the cost proportional to alignment length. Initial highway pavement cost and costs of miscellaneous highway facilities for vehicle operation (such as guardrails, lighting poles, and medians installation costs) may be included in this category.

$$C_L = u_P L_N W_P + u_L L_N \qquad (5.6)$$

$$W_P \leq W_N \qquad (5.7)$$

$$W_N = W_L + W_S \qquad (5.8)$$

where, u_P = unit pavement cost in \$/m²

u_L = unit length-dependent cost except pavement cost
L_N = the length of a new highway alignment generated
W_P = width of paved portion of a new highway; $w_P \leq w_N$
W_N = width of a new highway
W_L = total travel lane width of the new highway
W_S = total shoulder width of the new highway
Other terms are defined previously. See, Table 5.1.

   In the optimization model, the length of a new highway alignment (L_N) is iteratively updated from the alignment generation process. The resulting alignment consists of tangent sections and curved sections; spiral transition curves coupled with circular curves may be added to the curved sections. The alignment length can be expressed as follows. Note that Eq. (5.9) is used for calculating the alignment length when one circular and two spiral transition curves are incorporated in a horizontal curve, while Eq. (5.10) is employed when only a circular curve is used.

Case 1: for horizontal circular curves with spiral transitions

$$L_N = \sum_{i=0}^{n} \sqrt{\left(x_{ST_i} - x_{TS_{i+1}}\right)^2 + \left(y_{ST_i} - y_{TS_{i+1}}\right)^2}$$

$$+ \sum_{i=0}^{n} \left[l_{ST_{i1}} + l_{ST_{i2}} + R_{H_i}\left(\theta_{PI_i} - 2\theta_{ST_i}\right)\right] \qquad (5.9)$$

Case 2: for horizontal circular curves without spiral transition

$$L_N = \sum_{i=0}^{n} \sqrt{\left(x_{CT_i} - x_{TC_{i+1}}\right)^2 + \left(y_{CT_i} - y_{TC_{i+1}}\right)^2}$$

$$+ \sum_{i=1}^{n} R_{H_i}\theta_{PI_i} \qquad (5.10)$$

where, (x_{CT_i}, y_{CT_i}) = coordinates of ith **CT** (circular-to-tangent) point; i.e., end of a circular curve

(x_{TC_i}, y_{TC_i}) = coordinates of ith **TC** (tangent-to-circular) point; i.e., beginning of a circular curve)

Other terms are defined previously (see Table 4.2).

5.4.1.2 *Right-of-way cost (C_R)*

The right-of-way cost (C_R) can be estimated by computing total cost of areas affected by the highway within its required right-of-way boundary. Thus, values of the affected land parcels as well as their actual areas needed for the right-of-way of the highway are necessary for the right-of-way cost estimation.

$$C_R = \sum_{k=0}^{n_{PC}} u_{v_k} A_k \qquad (5.11)$$

where, u_{v_k} = unit cost (property value) of the kth land parcel affected by the highway alignment

Other terms are defined previously. See, Table 5.1.

Note that the right-of-way cost function shown in Eq. (5.11) is just land-acquisition cost required for the right-of-way of the new highway. Besides the land-acquisition cost, reduction of property value due to a nearby high-way as well as the usability of the remaining lands or site improvements cost due to the new highway construction may be added for more precise right-of-way cost estimation.

5.4.1.3 Earthwork cost (C_E)

The earthwork cost (C_E) is calculated based on (i) terrain profile (i.e., ground elevation) of the study area and (ii) road heights at each major break point (e.g., every 100 foot station) in the terrain surface along its vertical alignment. The ground elevation of the study area should be provided as a model input, and the vertical alignment is generated from the model. The earthwork cost function is formulated based on the average end area method, and C_H is the cost of moving earth between adjacent cut and fill sections to balance overall earthwork volume. The unit-cut and unit-fill costs and earth shrinkage factor are user-specifiable. A detailed description of Eq. (5.12) may be found in earlier publications by the authors (e.g., Jha and Schonfeld, 2003).

$$C_E = C_H + \frac{1}{2} \sum_{i=1}^{n_E} [\omega_0 u_{c_i} s_r A_{c_i} L_{E_i} + \omega_1 u_{f_i} A_{f_i} L_{E_i}$$

$$+ \omega_0 (u_{c_i} s_r A_{tc_i} + u_{f_i} A_{tf_i}) L_{E_i}] \qquad (5.12)$$

$$\omega_0 + \omega_1 + \omega_2 = 1 \qquad (5.13)$$

where, $\omega_0, \omega_1, \omega_2$ = binary integers used in earthwork cost computation; $\omega_0, \omega_1, \omega_2 = 0$ or 1; $\omega_0 = 1$ for a cut section, $\omega_1 = 1$ for a fill section, and $\omega_2 = 1$ for a transition section;

A_c, A_f, A_{tc}, A_{tf} = cross-sectional areas under cut, fill, transitional cut, and transitional fill conditions
   n_E = total number of highway sections for earthwork volume calculation
   u_c = unit cot cost
   u_f = unit fill cost
   s_r = earth shrinkage or swell factor
   L_E = length of highway section for earthwork volume calculation
   Other terms are defined previously.

5.4.1.4 *Structure cost* (C_S)

The costs required for highway structures (C_S) are also sensitive and dominating in highway construction, and thus they should also be included in the objective function. Many highway structures are also associated with the construction of a new highway. These may include bridges for crossing the rivers or valleys and cross-structures for intersecting existing highways. Since the costs required for those structures are sensitive and dominating in highway construction, they should also be included in the total agency cost. In the model, three types of interchanges (clover, diamond, and trumpet) are used to represent structures for crossing existing roadways. Costs required for each structure consist of pavement cost, right-of-way cost, earthwork cost, and small bridge cost for grade separation of the existing road at the interchange. For estimating the interchange pavement cost, right-of-way cost, and earthwork cost, those for the highway basic segment can be used. Equation (5.15) can be utilized for estimating the small bridge cost. The at-grade intersection cost also consists of pavement cost, right-of-way cost, and earthwork cost. Note that the intersection right-of-way cost can be roughly estimated with Eq. (5.11).

$$C_S = C_{SB} + C_{SIC} + C_{SIS} + C_{ST} \tag{5.14}$$

$$C_{SB} = \alpha_0 + \alpha_1 l_B w_B \tag{5.15}$$

$$C_{SIC} = C_{PIC} + C_{RIC} + C_{EIC} + C_{BIC} \tag{5.16}$$

$$C_{SIS} = C_{PIS} + C_{RIS} + C_{EIS} \tag{5.17}$$

$$C_{PIS} = u_P A_{PIS} \tag{5.18}$$

$$C_{EIS} = u_f E_{VIS} \tag{5.19}$$

$$C_{ST} = C_{ET} + C_{aT} \tag{5.20}$$

$$C_{ET} = \pi u_T l_T (R_T)^2 \tag{5.21}$$

$$C_{aT} = \gamma_0 + \gamma_1 l_T + \gamma_2 (l_T)^2 \tag{5.22}$$

where, C_{SB} = small bridge cost for grade separation

$C_{SIC} =$	interchange cost
$C_{SIS} =$	intersection cost
$C_{ST} =$	tunnel cost
$\alpha_0 \alpha_1 =$	coefficients used in bridge cost computation
$l_B, w_B =$	length and width of a small highway bridge for grade separation
$C_{BIC} =$	small bridge cost for grade separation of the existing road at the interchange
$C_{EIC} =$	interchange earthwork cost
$C_{pIC} =$	interchange pavement cost
$C_{RIC} =$	interchange right-of-way cost
$C_{EIS} =$	intersection earthwork cost
$C_{pIS} =$	intersection pavement cost
$C_{RIS} =$	intersection right-of-way cost
$A_{pIS} =$	intersection pavement area
$C_{aT} =$	additional tunnel cost which includes cost for ventilation and lighting
$C_{ET} =$	tunnel earthwork cost
$E_{VIS} =$	earthwork volume required for the at-grade intersection
$l_T =$	tunnel length
$R_T =$	tunnel radius
$u_T =$	unit tunnel earthwork cost
$\gamma_0, \gamma_1, \gamma_2 =$	coefficients used in additional tunnel cost computation

Other terms are defined previously.

Tunnel cost consists of earthwork cost and additional tunnel cost which accounts for the cost of ventilation and lighting along the tunnel. The tunnel earthwork cost can be roughly estimated based on the radius and length of tunnel, while a quadratic function of tunnel length is used for estimating the additional tunnel cost (Kim *et al.*, 2007). For estimating cost of a large scale bridge for crossing rivers or valleys, linear cost functions based on the bridge span length and pier height may be used instead of Eq. (5.15).

5.4.1.5 *Maintenance cost* (C_M)

To realistically evaluate the highway maintenance cost (C_M), the entire section of a new highway is subdivided into two sub-categories — 1) highway basic segments and 2) highway bridges — since different sources of unit costs are used for estimating their maintenance costs in literature. The highway maintenance cost (C_M) function can be expressed as follows:

$$C_M = C_{MH} + C_{MB} \tag{5.23}$$

where, C_{MH} = present value of the maintenance cost for highway basic segments C_{MB} = present value of bridge maintenance cost.

The maintenance cost for highway basic segments (C_M^H) is normally length-dependent; that is, it is proportional to the length of the road segment. Thus, given the length of the highway segment and its unit maintenance cost (normally $ per unit distance per year), C_M^H cost can be expressed as in Eq. (5.24). This cost normally includes expense of routine highway maintenance required annually, such as repair of roadway pavement, guardrail, and median and drainage. Road resurfacing and rehabilitation costs may be included in the maintenance cost if the project evaluation period exceeds the highway's design-life. On the other hand, bridge maintenance cost (C_M^B) is incurred as a result of annual inspection, annual maintenance, and periodic rehabilitation. This cost can be estimated with annual bridge operation cost and the number of bridges needed for the new highway facility. Its cost function can be expressed with a unit maintenance cost (u_M^B) as shown in Eq. (5.25).

$$C_{MH} = \left[L_N - \sum_{i=1}^{n_B} l_{B_i} \right] \left[u_{MH} \sum_{k=1}^{n_y} \left(\frac{1}{1+\rho} \right)^k \right] \tag{5.24}$$

$$C_{MB} = \sum_{i=1}^{n_B} \left[u_{MB} l_{B_i} \sum_{k=1}^{n_y} \left(\frac{1}{1+\rho} \right)^k \right] \tag{5.25}$$

where, n_B = total number of bridges in the highway alignment generated

n_y = analysis period or design life of a road
ρ = annual interest rate
u_{MH} = unit maintenance cost for highway basic segments
u_{MB} = unit bridge maintenance cost

Other terms are defined previously.

In Eq. (5.24), u_M^H normally includes costs of routine highway mainte-nance required annually, such as repair of roadway pavement, guardrail, and median and drainage. Road resurfacing and rehabilitation costs may be included in the maintenance cost if the project evaluation period exceeds the highway's design-life. The value of u_M^H can be found from many studies on the subject. For instance, Safronetz and Sparks (2003) used \$0.888/ft/yr (\$2,915/km/yr) for estimating the road maintenance cost in their highway management model, and Christian and Newton (1999) pro-posed \$0.914/ft/yr (\$3,000 /km/yr) for the unit highway maintenance cost.

5.4.2   User Cost

The user cost for a new highway development includes: travel time cost (C_T), vehicle operation cost (C_V), accident cost (C_A). Mathematical for-mulations of the three user cost components are discussed in the next sub-sections.

5.4.2.1   *Travel time cost* (C_T)

The travel time cost (C_T) is calculated based on the amount of time spent for traveling and the drivers' perceived value of time. In the model, it is assumed that two types of user classes (auto and truck) operate on the new highway, and they have different values of travel time for different trip purposes. Note that a detailed trip purpose factor for each user class (such as, home-based work and non-home-based trips) is not considered in the model, since it may not be available in the initial stage of the highway project due to time and/or money constraints. Vehicle occupancy information is also used for evaluating travel time cost of highway users more precisely.

$$C_T = [\mathbf{x} \cdot \mathbf{t} \cdot \mathbf{H}] \cdot [\mathbf{v} \cdot \mathbf{T} \cdot \mathbf{o}] \left(\frac{e^{(r_t - \rho)n_y} - 1}{r_t - \rho} \right) \tag{5.26}$$

where, $\mathbf{x}$ = a vector of average traffic volume in different time frames; $\mathbf{x} = [x_{AM}, x_{PM}, x_{OFF}]$

$\quad \mathbf{t}$ = a vector of average travel time in different time frames; $\mathbf{t} = [t_{AM}, t_{PM}, t_{OFF}]$

$\quad \mathbf{H}$ = a vector of duration of different time frames per year; $\mathbf{H} = [H_{TAM}, H_{TPM}, H_{TOFF}]$

$\mathbf{v}$ = a vector of unit travel time values for different modes

$\mathbf{T}$ = traffic composition vector representing different modes on the new highway

$\mathbf{o}$ = a vector of average vehicle occupancies for different modes

r_t = annual traffic growth rate

Other terms are defined previously.

5.4.2.2 *Vehicle operating cost (C_V)*

Another user cost component considered in the model is the vehicle operating cost (C_V) that can be directly perceived by the highway users as an out-of-pocket expense incurred while operating vehicles. This may include fuel consumption, maintenance, tire wear, and vehicle depreciation costs. However, since the vehicle depreciation cost is not sensitive to different highway alternatives, only the first three items, which are the most dominating and sensitive ones, are considered in the model. Generally, the vehicle operation cost can be calculated based on a per vehicle-mile basis, and thus length of highway alignment as well as traffic information (such as traffic flow, travel time, and speed) are necessary for estimating it.

$$C_V = L_N \left[\mathbf{x} \cdot \mathbf{H}\right] \cdot \left[\mathbf{u} \cdot \mathbf{T}\right] \left(\frac{e^{(r_t - \rho)n_y} - 1}{r_t - \rho}\right) \tag{5.27}$$

$$u_j = p_{F_j} f_j + m_j; \quad u_j \in \mathbf{u} \tag{5.28}$$

where, L_N = the length of a new highway alignment generated

$\mathbf{u}$ = a vector of unit vehicle operation costs for different type of modes

p_{F_j} = fuel price of jth mode operated on the new highway

f_j = fuel consumption rate of jth mode operated on the new highway

m_j = vehicle maintenance cost of jth mode

Other terms are defined previously.

5.4.2.3 *Accident cost (C_A)*

Estimating highway accident cost (C_A) is relatively difficult since accidents are caused by combinations of various factors (such as traffic volume, highway geometry, and driving conditions of users operating on a highway). The accident cost can be estimated with unit accident cost (e.g., $/accident) and the number of accidents predicted from an accident regression analysis. In the highway accident literature, it is obvious that highway geometric design elements affect road collisions. For instance, a sharp horizontal curve with insufficient tangent approach may cause a significant safety problem (Glennon *et al.*, 1985). Thus, the new highway must satisfy highway design standards, and collision effects of various highway alternatives with different geometric design elements should be evaluated in the alignment optimization process.

$$C_A = u_A F_A \left(\frac{e^{(r_t - \rho)n_y} - 1}{r_t - \rho} \right) \qquad (5.29)$$

where, u_A = unit accident cost

F_A = average accidents (frequency) estimated
Other terms are defined previously.

Note that the HAO model is designed with a modular structure in which various evaluation components can be easily replaced without changing the rest of the model structure. Thus, any available accident prediction relations or models can be incorporated in the model for estimating the accident frequency of new highways. Note that a variety of accident prediction models has been developed for predicting accidents on highway segments. Among them, theoretical models by Vogt and Bared (1998), Zeger *et al.* (1992), and Chatterjee *et al.* (2003) can be adopted in the model to predict F_A.

5.4.3   Penalty and Environmental Costs

5.4.3.1   *Penalty cost* (C_P)

Various types of environmental and socio-economic areas may be included in the study area of a new highway construction. These are, for example, wetland, wild-life refuge, and residential areas. Alignment's impacts on such land-use types should be as minimal as possible and if any, special care should be taken to replace and restore them. In the model, GIS maps containing various geographic entities are provided as a model input, and a highway alignment under evaluation is overlaid on the GIS maps for estimating its impact to the study area. Thus, the fractions of affected land parcels needed for the alignment is computed. If the area of the land parcel affected by the highway alignment exceeds its pre-defined maximum allowable limit (defined as *MaxA*), a penalty is applied to the excess area. Equation (5.31) is a soft penalty function, which can be used for computing the penalty associated with environmentally sensitive areas taken by the new highway alignment. Note that Eq. (5.30) is included in the objective function, Eq. (5.1) to smoothly guide the search in the optimization process.

A soft penalty function is also used in a case where a highway alignment generated from the GA-based HAO model violates highway design constraints. Equations (5.34) to (5.39) are penalty functions, used in the model, to handle horizontal and vertical alignments that violate the following design constraints:

- Minimum horizontal curve radius (R_{H_m})
- Minimum horizontal sight distance (S_{H_m})
- Minimum length of spiral transition curve (S_{T_m})
- Minimum length of vertical curve (L_{V_m})
- Minimum length of vertical sight distance (S_{V_m})
- Maximum allowable gradient (g_{max})

$$C_P = C_{PE} + C_{PD^H} + C_{PD^V} \tag{5.30}$$

$$C_{PE} = \sum_{k=1}^{n_{PC}} [\beta_{E^0} + \beta_{E^1} I_{PE_k} (A_k - MaxA_k)^{\beta_{E^2}}], \quad \text{only if } A_k > MaxA_k \tag{5.31}$$

$$C_{PD^H} = \sum_{i=1}^{n_{HC}} [C_{PD^{HR}} + C_{PD^{HS}} + C_{PD^{ST}}] \tag{5.32}$$

$$C_{PD}^V = \sum_{i=1}^{n_{VC}} [C_{PD^{VL}} + C_{PD^{VS}} + C_{PD^{VG}}] \tag{5.33}$$

$$C_{PD^{HR}} = \sum_{i=1}^{n_{HC}} [\beta_{HR^0} + \beta_{HR^1} (R_{H_m} - R_{H_i})^{\beta_{HR^2}}], \quad \text{only if } R_{H_i} < R_{H_m} \tag{5.34}$$

$$C_{PDHS} = \sum_{i=1}^{n_{HC}} [\beta_{HS^0} + \beta_{HS^1}(S_{H_m} - S_{H_i})^{\beta_{HS^2}}], \quad \text{only if } S_{H_i} > S_{H_m}$$

(5.35)

$$C_{PDST} = \sum_{i=1}^{n_{HC}} [\beta_{ST^0} + \beta_{ST^1}(S_{T_m} - S_{T_i})^{\beta_{ST^2}}], \quad \text{only if } S_{T_i} < S_{T_m}$$

(5.36)

$$C_{PDVL} = \sum_{i=1}^{n_{VC}} [\beta_{VL^0} + \beta_{VL^1}(L_{V_m} - L_{V_i})^{\beta_{VL^2}}], \quad \text{only if } L_{V_i} < L_{V_m}$$

(5.37)

$$C_{PDVS} = \sum_{i=1}^{n_{VC}} [\beta_{VS^0} + \beta_{VS^1}(S_{V_m} - S_{V_i})^{\beta_{VS^2}}], \quad \text{only if } S_{V_i} < S_{V_m}$$

(5.38)

$$C_{PDVG} = \sum_{i=1}^{n_{VC}} [\beta_{VG^0} + \beta_{VG^1}(g_i - g_{max})^{\beta_{VG^2}}], \quad \text{only if } |g_i| < g_{max}$$

(5.39)

where, C_{PE} = penalty associated with environmentally sensitive areas taken by the new highway

C_{PDH} =	penalty cost for violating design constraints of horizontal alignments
C_{PDHR} =	penalty cost for violating the minimum horizontal curve radius
C_{PDHS} =	penalty cost for violating design constraints of horizontal alignments
C_{PDST} =	penalty cost for violating the minimum length of spiral transition curve
C_{PDV} =	penalty cost for violating design constraints of vertical alignments
C_{PDVL} =	penalty cost for violating the minimum length of vertical curve
C_{PDVS} =	penalty cost for violating the minimum vertical sight distance

$C_{PD^{VG}} =$	penalty cost for violating the maximum allowable gradient
$n_{HC} =$	total number of horizontal curve sections of the highway alignment generated
$n_{VC} =$	total number of vertical curve sections of the highway alignment generated
$R_{H_i} =$	horizontal curve radius at the ith horizontal curve section
$S_{H_i} =$	horizontal sight distance at the ith horizontal curve section
$S_{T_i} =$	spiral transition curve length at the ith horizontal curve section
$L_{V_i} =$	vertical curve length at the ith vertical curve section
$S_{V_i} =$	vertical sight distance at the ith vertical curve section
$g_i =$	forward or back tangent grade at the ith vertical curve section
$I_{PEk} =$	a dummy variable indicating if the kth parcel is the environmentally sensitive area
$\beta_{E0}, \beta_{E1}, \beta_{E2} =$	coefficients used in computing C_{PE}
$\beta_{HR0}, \beta_{HR1}, \beta_{HR2} =$	coefficients used in computing $C_{PD^{HR}}$
$\beta_{HS0}, \beta_{HS1}, \beta_{HS2} =$	coefficients used in computing $C_{PD^{HS}}$
$\beta_{ST0}, \beta_{ST1}, \beta_{ST2} =$	coefficients used in computing $C_{PD^{ST}}$
$\beta_{VG0}, \beta_{VG1}, \beta_{VG2} =$	coefficients used in computing $C_{PD^{VG}}$
$\beta_{VL0}, \beta_{VL1}, \beta_{VL2} =$	coefficients used in computing $C_{PD^{VL}}$
$\beta_{VS0}, \beta_{VS1}, \beta_{VS2} =$	coefficients used in computing $C_{PD^{VS}}$

Other terms are defined previously.

5.4.3.2 *Environmental cost* (C_{EN})

Estimating highway environmental cost (C_{EN}) is complex as many factors (such as traffic, traffic mix, and rainfall intensity and frequency) are involved in the cost, besides highway profiles (e.g., vertical curves, grades, and length) and geographical locations where the highway is located. The environmental cost includes air, noise, and water pollution costs associated with highway development. A unit environmental cost (u_E) per vehicle

mile traveled is used in the model to estimate environmental impacts due
to air, noise, and water pollution. Taking into account of a life-cycle cost
concept throughout a design-life of a highway, the environmental cost can
be formulated as Eq. (5.40).

$$C_{EN} = \frac{1}{5280} u_E \, [\mathbf{x} \cdot \mathbf{H}] \left(\frac{e^{(r_t - \rho)n_y} - 1}{r_t - \rho} \right) \tag{5.40}$$

where, u_E = unit environmental cost per vehicle mile traveled (VMT)
    Other terms are defined previously.

5.4.4 Life-cycle Cost

Road maintenance cost occurs throughout the entire design-life of the road.
The user cost also persists over the system design life. Traffic demand fluc-
tuates over daily, monthly, and yearly cycles, and tends to increase over
the system design life. Thus, these costs should be estimated as a life-cycle
cost, by being discounted over the estimated system life at an appropriate
interest rate and traffic growth rate. Equations (5.23) to (5.29) show the formulations used in the model for estimating the present value of those costs.
Environmental costs (C_{EN}), which include air, noise, and water pollution
costs triggered along the highway, may also be analyzed over the system
design life because they are also affected by traffic characteristics as well
as road condition.

Chapter 6

Constraint Handling for Evolutionary Algorithms

Evolutionary algorithms (EAs),[1] which constitute a subfield of artificial intelligence (AI), have been developed for solving many complex constraint optimization problems. GAs are one of the most popularly used evolutionary algorithms that have many successful applications to complex optimization problems (e.g., project scheduling optimization (Wang and Schonfeld, 2007), space structures optimization (Adeli and Cheng, 1993), and steel structures optimization (Sarma and Adeli, 2000)). If GAs are applied to solve the complex problems, the method of handling infeasible solutions (generated from the algorithms) is very important to the effectiveness of the solution search process. This problem arises because solution search techniques used in GAs (such as reproduction, mutation, and recombination) are normally "blind" to the constraints, and thus it is possible for the GAs to generate a solution which does not satisfy the requirements of the problems (Michalewicz and Michalewicz, 1995).

Various constraint handling techniques used in GAs are reviewed in Craenen *et al.* (2003) and Coello (2002). They are normally classified into two groups: (i) direct methods and (ii) indirect methods. According to the authors, "direct constraint handling" means that violating constraints is not reflected in the optimization objectives (i.e., fitness or objective function) so that there is no bias towards solutions satisfying them. On the contrary, the objective function includes penalties for constraint violation in case of the "indirect constraint handling" approaches. Typical approaches categorized in the two cases are summarized in Table 6.1, and their general advantages and disadvantages are presented in Table 6.2. Additionall y, Table 6.3 shows several studies associated with each approach.

---

[1] "EAs are a class of direct, probabilistic search and optimization algorithms gleaned from the model of organic evolution." (Back, 1996)

Table 6.1 Typical constraint handling methods used in evolutionary algorithms

Control strategy	Approaches
Direct constraint handling	— Eliminating infeasible solutions — Repairing infeasible solutions — Preserving feasibility by special operators — Decoding (i.e., transforming the search space) — Locating the boundary of the feasible region
Indirect constraint handling	— Assigning penalty to objective of infeasible solutions

Source: Craenen *et al.* (2003) and Coello (2002).

Table 6.2 General advantages and disadvantages of constraint handling methods

Control strategy	Advantages	Disadvantages
Direct constraint handling	— May perform very well with significant improvement on computation efficiency — Might naturally accommodate existing heuristics	— Is usually problem-dependent — Might be difficult to design a method for a given problem
Indirect constraint handling	— Is easily applicable (i.e., is not problem-dependent) — Reduces complex ones to simple problems — Allows user preferences by weights	— Requires many penalty parameters — Requires prior knowledge of degree of constraint violation — Does not contribute computational efficiency (evaluate all solutions)

Source: Craenen *et al.* (2003) and Coello (2002).

Table 6.3 Constraint handling methods used in evolutionary algorithms

	Approach	References
Direct constraint handling	Elimination	Michalewicz and Xiao (1995); etc.
	Repairing	Liepins and Vose (1990); Liepins and Potter (1991); Michalewicz and Janikow (1991); Nakano (1991); Muhlenbein (1992); Orvosh and Davis (1993 and 1994); Le Riche and Haftka (1994); Michalewicz and Xiao (1995); Tate and Smith (1995); Xiao *et al.* (1996 and 1997); Steele *et al.* (1998); etc.
	Preserving	Davis (1991); Michalewicz and Janikow (1991); Michalewicz *et al.* (1991); Michalewicz (1996); Whitley (2000); etc.
	Decoding	Palmer and Kershenbaum (1994); Dasgupta and Michalewicz (1997); Kim and Husbands (1997 and 1998); Kowalczyk (1997); Koziel and Michalewicz (1998); etc.
	Locating boundary of feasible regions	Schoenauer and Michalewicz (1996 and 1998); etc.
Indirect-constraint handling (Penalty method)	Death penalty	Schwefel (1981); Back *et al.* (1991); etc.
	Static penalty	Richardson *et al.* (1989); Goldberg (1989); Back and Khuri (1994); Homaifar *et al.* (1994); Huang *et al.* (1994); Olsen (1994); Thangiah (1995); Le Riche *et al.* (1995); Morales and Quezada (1998); etc.
	Dynamic penalty	Joines and Houck (1994); Michalewicz (1995); Kazarlis and Petridis (1998); etc.
	Annealing penalty	Michalewicz and Attia (1994); Carlson and Shonkwiler (1998); etc.
	Adaptive penalty	Hadj-Alouane and Bean (1992); Smith and Tate (1995); Yokota *et al.* (1996); Gen and Cheng (1996); Eiben and Hauw (1998); etc.
	Co-evolutionary penalty	Coello (2000)

6.1   Direct Constraint Handling

6.1.1   Elimination Method

Elimination methods, which are also known as abortion methods (Michalewicz and Michalewicz, 1995),[2] are employed to remove infeasible solutions from the population. Such methods are aimed for avoiding evaluation of fitness values of infeasible solutions which are possibly generated from a GA. Thus, no infeasible solutions are allowed to be in the population although they are generated from genetic operators embedded in the GA. Elimination is a popular option in many GA applications. However, it has two major drawbacks. First, the elimination method does not allow to the search any chance to traverse on infeasible part of the search space. However, as stated in Michalewicz and Michalewicz (1995), "quite often the system can reach the optimal solutions by crossing an infeasible region especially in non-convex feasible search spaces". Thus, only using elimination methods as constraint handling methods of a GA application should be prohibited. Instead, some combination of other methods (such as repairing, decoding or penalty approaches) would be preferable. Another drawback of the method is that it is usually problem-dependent so that a specific elimination procedure is needed for every particular problem.

6.1.2   Repairing Method

The repairing method is another popular method used in EAs. The main concept of this method is a combination of learning and evolution processes. Through the iterative learning process (e.g., local search for the closest feasible solution), an infeasible solution can be repaired with improved objective value. Note that this method allows presence of infeasible solutions in the population on the contrary to the elimination method. This allows an EA to search infeasible parts of the search space. When an infeasible solution can be easily repaired into a feasible solution, using repairing algorithms may be a good choice for an efficient GA. However, this is not always possible, and in some cases repair algorithms may cause a deterioration in

---

[2]Another well-known classification scheme of the constraint handling techniques for EAs is that of Michalewicz and Michalewicz (1995). They distinguish "pro-life" and "pro-choice" approaches, where "pro-life" methods allow the presence of infeasible solutions in the population, while "pro-choice" approaches disallow it. "Pro-life" covers penalty and repairing methods, while elimination (abortion), decoding, and preserving methods can be classified into "pro-choice" category.

the search process of GAs. Furthermore, for some complex constraint optimization problems (e.g., scheduling and timetable problems), the process of repairing infeasible solutions might be as complex as solving the original problem itself (Michalewicz and Michalewicz, 1995). Another shortcoming of the repair approaches is that there are no standard heuristics for design of repair algorithms since they are also problem-dependent, like elimination methods.

6.1.3 Preserving Method

The main concept of this method is to maintain the feasibility of solutions in the population. Many special operators have been developed for using them as the preserving method. These are partial-mapped crossover (PMX), order crossover (OX), position-based crossover (PBX), order-based crossover (OBX), and edge recombination crossover (ERX). They are designed for prohibiting an illogical sequence of genes in offspring which may result from permutation representation with the traditional one-point or two-point crossovers. Note that the preserving approach also has some limitations in its application. The use of the special operators is useful only for the specific application for which they were designed (e.g., project scheduling problem and traveling salesman problem). Application of those operators to a GA in which the genes are represented by real numbers (e.g., XYZ coordinates) rather than binary digits may not be appropriate (this shows that the preserving method is also problem-dependent). In addition, the preserving method requires an initial feasible population, which can pose a hard problem by itself (Craenen *et al.*, 2003). More detailed discussion of the special operators is provided in Goldberg (1989); Gen and Cheng (1996); and Michalewicz (1996).

6.1.4 Decoding Method

The main idea of the decoding method is to transform the original problem (domain of original search space) into another form that is easier to optimize by EAs. This method does not allow generation of infeasible solutions. For instance, a sequence of items for the knapsack problem can be interpreted as a sequence of binary digits ("0" or "1") with an instruction "take an item if possible". By simplifying the problem with an effective decoding method, computation time of GAs can be significantly reduced. Several conditions that must be satisfied when using the decoding method are proposed by Dasgupta and Michalewicz (1997): "1) for each feasible solution s there must be a decoded solution d, 2) each decoded solution d must correspond to a feasible solution s, and 3) all feasible solutions should be represented by the same number of decoding d. Additionally, it is reasonable to request that 4) the transformation Tr is computationally fast, and 5) it has a locality feature in the sense that small changes in the decoded solution result in small changes in the solution itself". Despite several advantages (see Koziel and Michalewicz, 1998), this method also has some shortcomings. Designing a decoding method for a given problem may be significantly difficult since this method is also problem-dependent; in addition, a transformed problem from a designed decoding method may require more computation time than that required in its original problem.

6.1.5 Locating the Boundary of the Feasible Regions

The main idea in this method is to search areas close to the boundary of the feasible region. According to Coello (2002), the idea was originally proposed in an Operation Research technique known as strategic oscillation (Glover, 1977), and has been used in some combinatorial and nonlinear optimization problems (Glover and Kochenberger, 1995). This approach has two basic components: (1) an initialization procedure that is designed for generating feasible solutions, and (2) genetic operators that are employed to explore the feasible region (Coello, 2002). Note that since the approach allows exploring feasible and infeasible regions close to the boundary, a penalty approach may be added to it. The main drawback is that the method is also highly problem-dependent. In addition, it may require complex computation since the feasible regions of the complex optimization problems are usually non-convex and irregular in form. Moreover, there may be several disjoint feasible regions in the problem. Despite such limitations, the approach may be quite efficient and generate good outcomes whenever they are well implemented.

6.2 Indirect Constraint Handling (Penalty Approaches)

The penalty method is the most common approach used in EAs (particularly in GAs community) for handling complex constraints. The main idea of this method is to transform a constrained optimization problem into an unconstrained one by adding a certain value (penalty) to the fitness function of the given problem based on the amount (number or severity) of constraint violation presented in a solution.

The penalty should be kept as low as possible, just above the limit below which infeasible solutions are optimal (this is called, the minimum penalty rule (Smith and Tate, 1993; Le Riche *et al.*, 1995)). This should be maintained because an optimization problem might become very difficult for a GA if the penalty is too high or too low. A large penalty discourages GAs exploration to the infeasible regions so as not to move to different feasible regions unless they are very close. On the other hand, if the penalty is not severe enough, then too large a region is searched and much of the search time will be spent exploring the infeasible region due to its negligible impact to the objective function (Smith and Coit, 1997). Such a minimum penalty rule seems simple. However, it is not easy to implement because in many complex problems for which GAs are intended, the exact location of the boundary between the feasible and infeasible regions is unknown (Coello, 2002).

Several variations of penalty functions have been developed for handling the infeasible solutions. These are death penalty, static penalty, dynamic penalty, adaptive penalty, annealing penalty, and co-evolutionary penalty. Yeniay (2005); Coello (2002); and Smith and Coit (1997) extensively reviewed these penalty methods, accounting for advantages and disadvantages of each method. According to their researches, the main problem of most methods is to set appropriate values of the penalty parameters. They suggest that parameter values should be specified based on researchers' good judgments through many experiments. Some famous penalty approaches, which are relatively easy to apply in the GA-based optimization problems, are discussed in the following.

6.2.1 Death Penalty

The main idea in this method is just to assign a high penalty (i.e., $+\infty$ for minimization problems) when a solution generated from a GA violates any constraint. Therefore, no further calculations are necessary to estimate the degree of infeasibility of the solution. The death penalty method is simple and popular. However, it can perform well only if the feasible search space is not disjointed and constitutes a large portion of the whole search space. In addition, if there are no feasible solutions in the initial population (which is normally generated at random), then the evolutionary process will not improve since all the solutions will have the same fitness value (i.e., $+\infty$ for minimization problems). Many studies have reported that the use of this method is not a good choice (Coello, 2002; Smith and Coit, 1997). However, it should be noted that the death penalty method can significantly improve the search process of a GA if it works with other efficient constraint handling methods (e.g., a repairing method).

6.2.2 Static Penalty

In this method, simple penalty functions are used to penalize infeasible solutions. The reason why this method is called static penalty function is that penalty parameters in the function are not dependent on the current generation number. Two variations on this simple penalty method; one is constant static penalty and the other is a metric-based penalty function. The former method is to assign constant penalty value (Pcs) based on the number of constraints that a solution violates regardless of severity of the violations. For instance, if a solution violates n constraints, then the penalty added to the objective function is nPcs. It should be noted that the constant penalty method is generally inferior to the second approach which is based on some distance metric from the feasible region (Goldberg, 1989; Richardson *et al.*, 1989). The second approach, which is a more common and more effective penalty method, is to use a soft penalty that includes a distance metric for each constraint, and adds the penalty which becomes more severe with distance from feasibility (Smith and Coit, 1997). A general formulation of this penalty function is as follows:

$$P(\mathbf{j}) = \sum_{i=1}^{m} \eta_i d_i^k \qquad (6.1)$$

$$d_i = \begin{cases} \delta_i g_i(\mathbf{j}), & for\ i = 1, \cdots, q \\ |h_i(\mathbf{j})|, & for\ i = q+1, \cdots, m \end{cases} \qquad (6.2)$$

where, $P(\mathbf{j})$ = penalty applied to solution $\mathbf{j}$;
  η_i = penalty coefficient corresponding to i^{th} constraint;

d_i = distance metric of ith constraint;

δ_i = dummy variable; $\delta_i = 1$, if i^{th} constraint is violated; $\delta_i = 0$, otherwise;

$\mathbf{j}$ = vector of decision variables;

$g(\mathbf{j})$, $h(\mathbf{j})$ = functions of inequality and equality constraints, respectively;

q = the number of inequality constraints;

$m - q$ = the number of equality constraints;

d_i is the distance metric of constraint i applied to solution $\mathbf{j}$ and k is user defined exponent (normally $k = 1$ or 2). η_i indicates the penalty coefficient corresponding to i^{th} constraint and must be estimated based on the relative scaling of the distance metrics of multiple constraints, on the difficulty of satisfying a constraint, and on the seriousness of constraint violations, or be determined experimentally (Smith and Coit, 1997).

6.2.3 Dynamic Penalty

In the dynamic penalty method, the penalty coefficient (η_i) is usually dependent on the current generation number. Normally the penalty function is defined in such a way that it increases over the successive generations (Coello, 2002). The main idea of this method is "allowing highly infeasible solutions early in the search process, while continually increasing the penalty imposed to eventually move the final solution to the feasible region" (Smith and Coit, 1997). A general formulation of a distance-based penalty method incorporating a dynamic aspect is as follows:

$$P(\mathbf{j}) = \sum_{i=1}^{m} S_i(t)d_i^k \qquad (6.3)$$

where, $S_i(t)$ = a monotonically non-decreasing function;

t = the number of solutions searched;

Other terms are defined previously.

The primary defect of this method is that it is very sensitive to value of $S_i(t)$, and thus may result in infeasible solutions at the end of evolution. Therefore, this method typically requires problem-specific tuning to perform well (Smith and Coit, 1997). There is no evidence that this dynamic method performs better than the static penalty method.

6.2.4   Adaptive Penalty

The main idea in this method is to reflect a feedback from the search process into a penalty function. Penalty parameters are updated for every generation according to information obtained from the population. There is no general form of this method since it is also highly problem-dependent. Compared to the methods described earlier, relatively few studies (Hadj-Alouane and Bean, 1992; Smith and Tate, 1995; Yokota *et al.*, 1996; Gen and Cheng, 1996; Eiben and Hauw, 1998) have used this penalty method. Readers may refer to Table 6.3 for studies on this subject. Note that some researchers classify this method as a dynamic penalty method.

6.3   Handling Infeasible Solutions of GA-based Highway Alignment Optimization

Among many constraint handling methods available for evolutionary algorithms, the most common methods used in GA applications are the indirect constraint handling techniques (which are also known as penalty methods) because they are easy to apply for many complex optimization problems by allowing user-specifiable penalty parameters in their functional forms. However, normally many parameters are used in the penalty methods, and they must be calibrated from many trials and errors based on the good judgments of researchers. Furthermore, since the penalty approaches work indirectly in the optimization process (i.e., by adding penalties to the objective function of the given problem based on the amount of constraint violations, and evaluating all solutions including the infeasible ones), computational burdens may often arise if the problem requires a complex and time-consuming evaluation procedure.

Other approaches categorized in the direct constraint handling methods also have shown several advantages although they are highly problem-dependent. They might perform well with GAs by significantly improving computational efficiency; furthermore, they might naturally accommodate existing heuristics whenever applicable. Thus, familiarity with properties of the given problem is very important in applying these approaches. Readers may refer to Appendix A for many studies employing the direct and/or indirect methods in the evolutionary algorithms.

There are many possibilities for using combinations of both the direct and indirect methods in GA-based complex optimization problems. In the GA-based HAO problem, various complex constraints are specified for: (i) highway design features (e.g., horizontal and vertical curvature constraints and gradient constraint) and (ii) geographical and environmental considerations (e.g., environmentally sensitive areas and outside the area of interest). Although the penalty methods can handle the solutions violating those constraints, some of them may be more efficiently controlled by a direct method. For instance, solution alignments which violate design constraints may be easily repaired with a simple modification process without evaluating their fitness with penalties. In addition, a feasible-boundary approach may be useful for handling solution alignments violating the geographical constraints as shown in Kang *et al.* (2007b) because good representation of such constraints can guide generating feasible solution alignments during the search process, while reducing model computation time.

In the HAO model, a GIS is used for the detailed right-of-way cost estimation and environmental impact analysis for the solution alignments generated from GAs, and thus every generated alignment requires massive processing of GIS data during the evaluation process. Therefore, incorporation of a direct method (before the GIS evaluation) is preferable to evaluating all solutions (including infeasible ones) with a traditional penalty approach in terms of efficient solution search process. It should be noted that a well-designed direct constraint handling method may be more applicable to any GAs-based complex optimization problem which requires a detailed evaluation procedure.

Chapter 7

Highway Alignment Optimization Through Feasible Gates

This chapter describes an effective constraint handling method (called feasible gate (FG)) developed for improving computation efficiency of the alignment optimization process. The method mainly aims to realistically represent complex geographical (spatial) constraints for the alignment optimization as well as to control solution alignments violating those constraints. A research motivation of the FG method is discussed in Section 7.1, and its methodologies applied for horizontal and vertical alignments are presented in Sections 7.2 and 7.3, respectively. Two example studies presented in Section 7.4 demonstrate the capability of the proposed method. Finally, Section 7.5 summarizes this chapter.

7.1 Research Motivation of Feasible Gates

As discussed in Section 5.4, various cost items are associated with highway location selection and alignment design process. Among them, the effects of alignments on environmentally sensitive areas are often regarded as the most attractive ones and complex effects in recent highway construction projects. User preferences including political issues may also be critical in selecting rights-of-way of the alignments. These factors are normally intangible and not easily estimated in monetary values; however, they may greatly reduce the alignment search problem by excluding many possibilities and requiring alignments to pass through some narrow "gates" or "corridors".

Until recently, the previously developed HAO model has relied only on a penalty approach to guide the search toward better solutions. It assigned penalties to the cost functions if the solution alignments violated the corresponding constraints, and eventually screened out the candidate

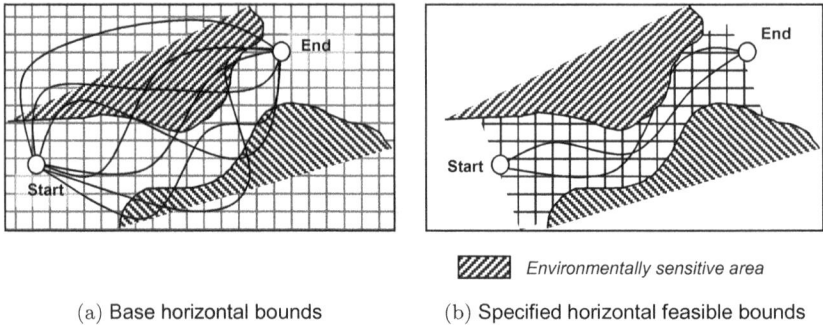

(a) Base horizontal bounds   (b) Specified horizontal feasible bounds

Figure 7.1 Bounded horizontal search space

solutions whose constraint violations were significant. However, finding the feasible solutions that satisfy geographical and environmental constraints, which are normally provided in undefined functional forms and are problem-dependent, with only such an indirect constraint handling method is computationally expensive. This is mainly because the model has to spend considerable time evaluating all generated solutions (including the infeasible solutions) with the penalty method.

As shown in Figure 7.1(a), many generated alignments may affect the existing environmentally sensitive areas since the search space is the entire area within the rectangular bounds. Such inefficiency is more severe if the sensitive areas are more complex so that the area of interest is also more complex or narrower. Obviously, the solution alignments that violate the sensitive areas cannot be the best solutions; furthermore, the detailed evaluation of each solution takes considerable time. Thus, a good representation of feasible area of interest is needed. An efficient use of the feasible search area can reduce computation time as well as improve solution qualities during the search process. In the model, the computational improvements are desirable since each candidate alignment requires massive processing of GIS data for its detailed evaluation.

In this chapter, feasible gate (FG) methods (for horizontal (HFG) and vertical (VFG) alignments) are proposed and named to ensure that complex preferences and environmental requirements are satisfied efficiently in the search process of the optimization model developed. The proposed

┌┬┬┬┬┐ *Feasible gates for vertical alignments*

(a) Base vertical bounds   (b) Specified vertical feasible bounds

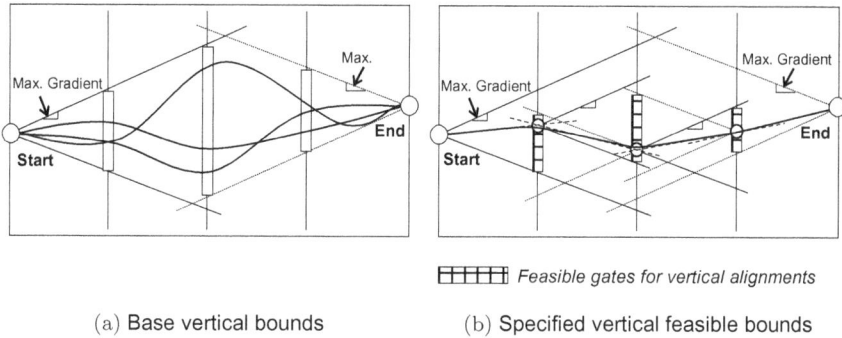

Figure 7.2 Bounded vertical search space

approaches are intended to avoid generating infeasible solutions that are outside the acceptable bounds and thus to focus the search on the feasible solutions. Figures 7.1(b) and 7.2(b) provide good insights into the proposed FG approaches for horizontal and vertical alignments, respectively. For both vertical and horizontal alignments, the points of intersections (PI's) are only generated randomly along the limited cutting planes orthogonal to the straight line connecting the start and end points, as shown in Figures 7.4 and 7.7 The key contribution of the feasible gate approaches is to limit the fraction of the cutting planes within which PI's for alignments can be generated, both horizontally and vertically. These limited "gates" are based on user preferences and environmental factors for horizontal alignments and on allowable gradients for vertical alignments, after adjustments to allow PI's outside feasible regions if the curved alignments at those PI's stay within feasible regions. By avoiding the generation and evaluation of many infeasible alignments outside the feasible regions, the search for optimized solutions is significantly accelerated. Particularly for horizontal alignments, since various spatial considerations apply, the preferred horizontal feasible gates may be quite complex, discontinuous, and significantly depend on the preferences of model users. Therefore, ways of dealing with the various user preferences and reflecting them in the solution search process are key issues to be resolved. It is relatively easier to ensure feasible gates for vertical alignments than for horizontal ones as vertical feasible ranges are usually bounded by maximum gradients.

7.2   Feasible Gates for Horizontal Alignments

A horizontal feasible gate (HFG) method is developed to realistically represent a complex horizontal search space in modeling highway alignments. In addition, since it requires interactive use of the spatial information in the study area, an input data preparation module (IDPM) is also developed. The IDPM is a customized GIS module designed for easy preparation of the HAO model inputs. With incorporation of the IDPM into the HFG-based approach, we now enable the model users to interactively specify their preferences (e.g., areas of interest) on given GIS maps and enhance the model solution quality and computation efficiency. The following subsections discuss how we define horizontal feasible bounds and gates to effectively optimize highway location and alignments. Note that notations used in the HFG method are presented in Table 7.1.

7.2.1   User-defined Horizontal Feasible Bounds

Figure 7.3 shows how the existing GIS maps and user's areas of interest are converted to the model-readable format through the IDPM. It is noted that digitized land use and property information (e.g., values and boundaries) maps are essential in using IDPM. Since GIS databases are widely used, some (even property maps) are available nowadays free or with some charges at the USGS (United States Geological Survey), ESRI (Environmental Systems Research Institute), and other websites of companies and local governments.

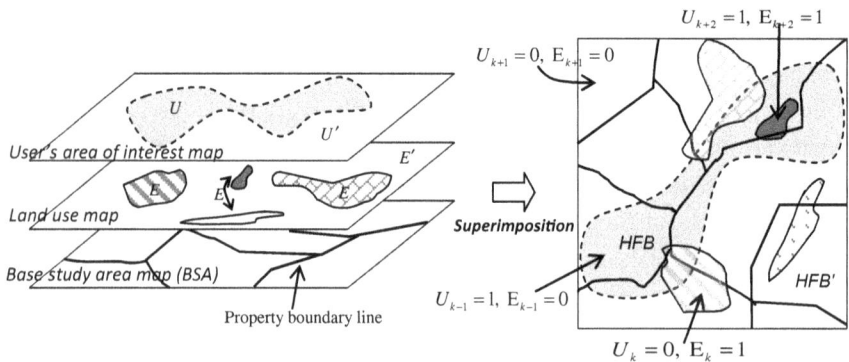

Figure 7.3   Setup of user-defined horizontal feasible bound with IDPM

Table 7.1 Notation used in Horizontal Feasible Gate (HFG) method

Notation	Descriptions
θ_{max}	maximum allowable deflection angle
θ_{vc}	an angle between a vertical cutting line and the X axis of a given map
$\beta_{HG_0}, \beta_{HG_1}, \beta_{HG_2}$	penalty parameters for user-defined geographical constraints
A_k^j	area of property piece k affected by alignment j
d_i	the coordinate of the intersection point at the ith vertical cutting line
d_{iU}, d_{iU}	the upper and lower bounds of d_i, respectively
D_{offset}	additional offset for horizontal feasible gate
E, E'	environmentally sensitive and insensitive areas, respectively
E_k	a dummy variable to indicate whether property piece k is inside E; ($E_k = 1$ if k is inside E; $E_k = 0$, otherwise)
F_i^q	q^{th} horizontal feasible gate for generating i^{th} PI_i; F_i^q can be defined with S_i^l, S_i^{l+1}, and D_{offset}; $q = 1, \ldots, m_i/2$
h, w	the height and width of the base study area (BSA), respectively
HFB, HFB'	horizontal feasible and untouchable areas, respectively
m_i	the total number of intersection points of $\overrightarrow{VC_i}$ with property pieces in the HFB
OB	the origin of the base study area (BSA); $OB = (x_{origin}, y_{origin})$
O_i	the origin of the ith vertical cutting line; $O_i = (x_{o_i}, y_{o_i})$
p_{S_i}	offset from the initial tangent to the PC of the shifted circle
R_{H_m}	minimum horizontal curve radius
R_{H_i}	horizontal curve radius at the ith horizontal curve section
S_i^l	i^{th} intersection point of $\overrightarrow{VC_l}$ with property pieces that are in the specified horizontal bounds (HFB), for $l = 1, \ldots, m_i$
$\overrightarrow{VC_l}$	i^{th} vertical cutting line for i^{th} horizontal PI
Start, End	horizontal start and end points of an alignment; $Start = (x_s, y_s)$; $End = (x_e, y_e)$
$\overline{SE}$	a line connecting the *Start* and *End*
U, U'	the area of interest and the area outside interest, respectively
U_k	a dummy variable to indicate whether property piece k is inside U; ($U_k = 1$ if k is inside U; $U_k = 0$, otherwise)

Let *BSA* be a base study area in which properties are spatially distributed in a rectangular space and k be a clipped property piece resulting from the superimposition of different map layers (refer to Figure 7.3). Additionally, let U be our area of interest, U' be the area outside it, and E and E' be environmentally sensitive and insensitive areas, respectively. Then U_k denotes whether property piece k is inside U ($U_k = 1$ if k is inside U; $U_k = 0$, otherwise). E_k indicates whether k is inside E ($E_k = 1$ if k is inside E; $E_k = 0$, otherwise). These variables are then used to represent horizontal feasible bounds (HFB) and untouchable areas (HFB').

$k \in HFB$, if $U_k = 1$ and $E_k = 0$ for all k in a project area

$k \in HFB'$, otherwise.

7.2.2   Representation of Horizontal Feasible Gates

Let $Start = (x_s, y_s)$ and $End = (x_e, y_e)$ be horizontal start and end points of a new alignment, and $\overline{SE}$ denotes the line connecting the two points. Vertical cutting lines $\overrightarrow{VC}$, which are perpendicular to $\overline{SE}$, are then introduced here to effectively search for horizontal PI's of a new highway alignment in a rectangular search space (see Figure 7.4). Note that a set of horizontal feasible gates (HFG) are described as a subset of the vertical cutting lines in the figure with specified-horizontal feasible bounds (HFB).

Suppose that we cut $\overline{SE}$ n times at equal distances between contiguous cuts and let $O_i = (x_{oi}, y_{oi})$ be the origin of the set of vertical cutting lines $\overrightarrow{VC_1}$, for all $i = 1, \ldots, n$. Then the coordinates of O_i can be expressed as:

$$\begin{bmatrix} x_{oi} \\ y_{oi} \end{bmatrix} = \begin{bmatrix} x_s \\ y_s \end{bmatrix} + \frac{i}{n+1} \begin{bmatrix} x_e - x_s \\ y_e - y_s \end{bmatrix} \tag{7.1}$$

Let θ_{vc} be the angle between the cutting line and the X axis of the given map coordinate system. Then the angle can be determined with Eq. (7.2).

$$\theta_{vc} = \tan^{-1}\left(\frac{y_n - y_0}{x_n - x_0} \right) + \frac{\pi}{2} \tag{7.2}$$

where, $0 < \theta_{vc} < \pi$.

We now let $OB = (x_{origin}, y_{origin})$ be the origin of the base study area (BSA) and h and w be height and width of the base study area, respectively. Then the i_{th} vertical cutting line vector, $\overrightarrow{VC_i}$ can be defined as the function of θ_{vc}, O_i, OB, h, and w. Additionally, let d_i be the coordinate of the intersection point at the i_{th} vertical cutting line, and d_{iU} and d_{iL} be its upper and lower bounds, respectively. Then, $\overrightarrow{VC_i}$, d_{iU}, and d_{iL} can be determined based on the following procedures. Note that all notations used in the procedures are listed in Table 7.1.

Case 1: If $\theta_{vc} = 0$ or π

$$d_{iU} = (x_{origin} + w) - x_{o_i} \tag{7.3}$$

$$d_{iL} = (x_{origin} - x_{o_i}) \tag{7.4}$$

$$\overrightarrow{VC_i} = \overline{(x_{origin}, y_{o_i})(x_{origin} + w, y_{o_i})} \tag{7.5}$$

Case 2: If $0 < \theta_{vc} < \pi/2$

$$d_{iU} = \min[(x_{origin} + w) - x_{o_i}/\cos\theta_{vc}, (y_{origin} + h) - y_{o_i}/\sin\theta_{vc}] \tag{7.6}$$

$$d_{iL} = \min[(x_{origin} - x_{o_i})/\cos\theta_{vc}, (y_{origin} - y_{o_i})/\sin\theta_{vc}] \tag{7.7}$$

$$\overrightarrow{VC_i} = \overline{(x_{o_i} + d_{iL}\cos\theta_{vc}, y_{o_i} + d_{iL}\sin\theta_{vc})(x_{o_i} + d_{iU}\cos\theta_{vc}, y_{o_i} + d_{iU}\sin\theta_{vc})} \tag{7.8}$$

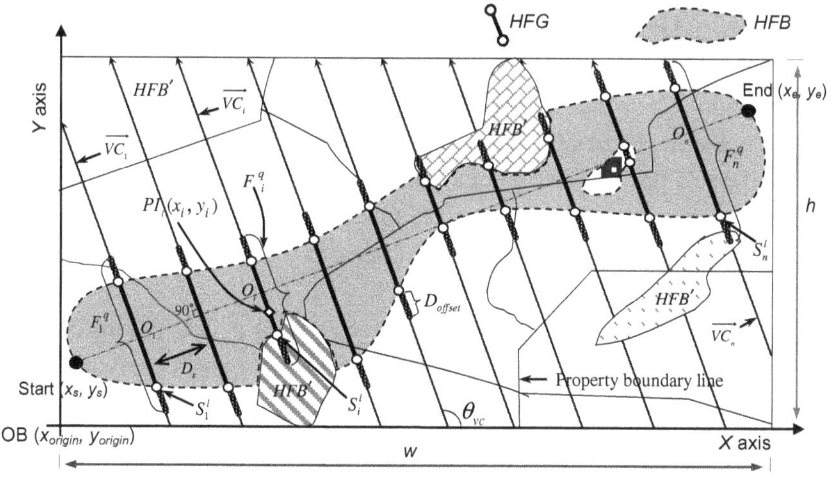

Figure 7.4 Representation of horizontal feasible gates

Case 3: If $\theta_{vc} = \pi/2$

$$d_{iU} = (y_{origin} + h) - y_{o_i} \qquad (7.9)$$

$$d_{iL} = (y_{origin} - y_{o_i}) \qquad (7.10)$$

$$\overrightarrow{VC_i} = \overline{(x_{o_i}, y_{origin})(x_{o_i}, y_{origin} + h)} \qquad (7.11)$$

Case 4: If $\pi/2 < \theta_{vc} < \pi$

$$d_{iU} = \min[(x_{origin} - x_{o_i})/\cos\theta_{vc}, (y_{origin} + h) - y_{o_i}/\sin\theta_{vc}] \qquad (7.12)$$

$$d_{iL} = \min[(x_{origin} + w) - x_{o_i}/\cos\theta_{vc}, (y_{origin} - y_{o_i})/\sin\theta_{vc}] \qquad (7.13)$$

$$\overrightarrow{VC_i} = \overline{(x_{o_i}+d_{iL}\cos\theta_{vc}, y_{o_i}+d_{iL}\sin\theta_{vc})(x_{o_i}+d_{iU}\cos\theta_{vc}, y_{o_i}+d_{iU}\sin\theta_{vc})}$$
$$(7.14)$$

Let PI_i be the horizontal point of intersection corresponding to i^{th} vertical cutting line vector $(\overrightarrow{VC_l})$ and S_i^l be the ith intersection point of $\overrightarrow{VC_l}$ with property pieces that are in the specified horizontal bounds (HFB) where $l = 1, \ldots, m_i$, and m_i is the total number of intersection points of $\overrightarrow{VC_i}$ with the property pieces in the HFB. Then, the qth horizontal feasible gate for PI_i, denoted as F_i^q can be determined by a line segment connecting the two consecutive intersection points (S_i^l and S_i^{l+1}) and an additional allowable offset (denoted by D_{offset}) where $q = 1, \ldots, m_i/2$. As shown in Figure 7.4, the set of horizontal feasible gates F_i^q for all i and all q outlines the specified horizontal feasible bound (HFB) and is designed to guide the model toward realistic horizontal alignments. The PI's are searched within the specified gates during the model's optimization process and determine the track of the horizontal alignments. Finally, the alignments resulting from the feasible PI's are obtained as candidates to be evaluated with detailed cost components embedded in the model.

Note that the additional allowable offset, which is denoted as D_{offset} in Figure 7.5, must be added to the horizontal feasible gates to avoid losing good candidate alignments since it is possible that excellent solutions run near borders between the specified feasible bounds and others. Figure 7.5(a) shows a limit of the horizontal feasible gates in a case where no allowable offset is provided. Some caution is required in determining the maximum deflection angle in order to fully use the HFG approach. D_{offset} becomes

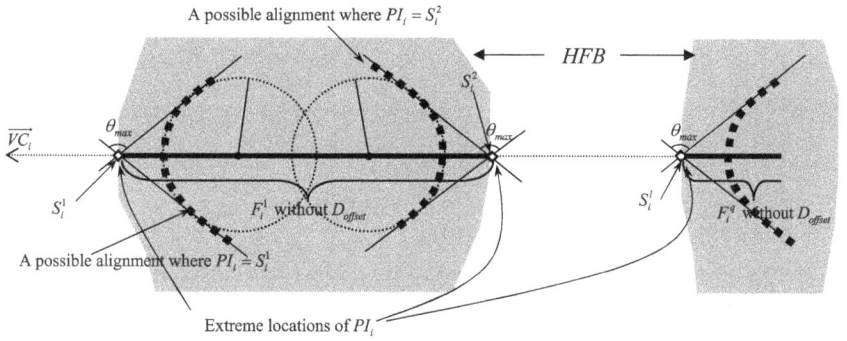

(a) **Extreme example alignments without** D_{offset}

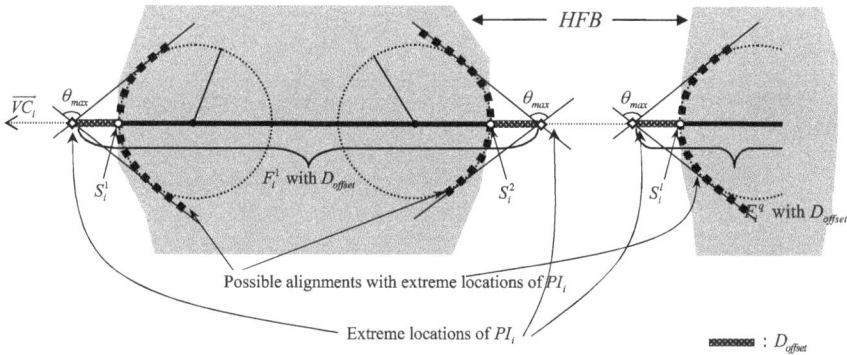

(b) **Extreme example alignments with** D_{offset}

Figure 7.5 Representation of allowable offsets near horizontal feasible gates

excessively long if θ_{max} is too large (e.g., more than $\pi/2$) and R_{H_i} is too long; thus, we hardly expect the benefit of the proposed HFG approach since the long allowable offset may cover the entire length of the vertical cutting line.

The minimum curve radius, R_{H_m} (i.e., the lower bound of R_{H_i}) is determined by the pre-specified design speed, maximum superelevation, and side friction factor. Thus, D_{offset} can be approximately estimated with Eqs. (7.15) and (7.16).

$$D_{offset} = R_{H_i} \left(\frac{1}{\cos(\theta_{max}/2)} - 1 \right)$$

if there is only a circular curve in the horizontal curved section

(7.15)

$$D_{offset} = (R_{H_i} + p_{S_i}) \sec(\theta_{max}/2) - R_{H_i}$$
if transition curves are added at both sides of the circular

(7.16)

where, R_{H_i} and θ_{max} are horizontal circular-curve radius and the maximum allowable; deflection angle at i^{th} *PI*, respectively;

Other terms are defined previously.

It should be noted in Eq. (7.16) that an additional parameter p_{S_i}, which is the offset from the initial tangent to the point of curvature of the shifted circle, needs to be defined to compute D_{offset} for a horizontal curved section with transition curves. Taking into account all these considerations, we summarize the feasible gate determination procedure in Section 7.2.3.

7.2.3 Horizontal Feasible Gate Determination Procedure

STEP 1: For $i = 1$ to n

Find S_i^l and m_i only if $\overrightarrow{VC_l}$ intersects k in *HFB* for all k in *HFB*

End

STEP 2: For $l = 1$ to m_i

Discard duplicated S_i^l

End

Update m_i

STEP 3: For $i = 1$ to n

For $q = 1$ to $m_i/2$

For $l = 1$ to m_i

$$F_i^q = \overline{S_i^l S_i^{l+1}}$$

Add D_{offset} at the both ends of F_i^q

End

End

End

7.2.4 User-defined Constraints for Guiding Feasible Alignments

In previous sections, horizontal feasible bounds (*HFB*) and feasible gates where PI's of horizontal alignments are generated are mathematically defined with illustration. It is noted, however, that the derived feasible gates do not always guarantee that feasible alignments are generated which satisfy

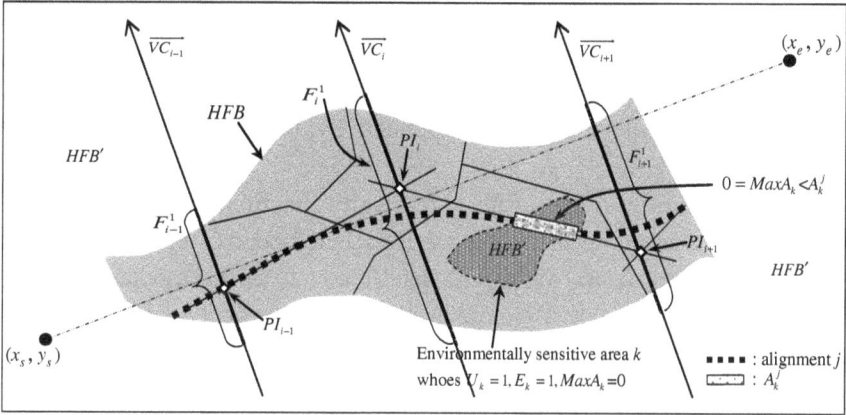

(a) An example alignment affecting the environmentally sensitive area

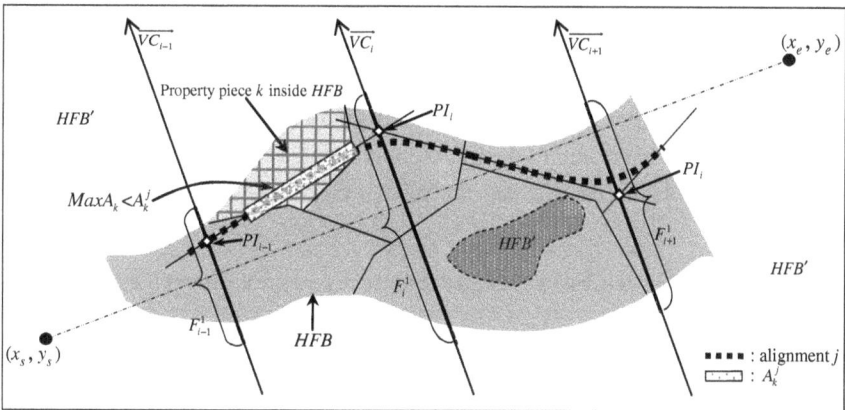

(b) An example alignment affecting user-specified property piece k with $MaxA_k < A_k^j$

Figure 7.6 Example alignments violating user-defined constraints

complex geographical constraints defined by the model users. For instance, solution alignments generated from the optimization model might still affect the untouchable areas (HFB') if they are surrounded by or in the middle of the feasible bounds as shown in Figure 7.6(a). In addition, the alignments might affect areas of property piece k by more than the allowable amounts specified by the model users, as shown in Figure 7.6(b).

To address such a problem, a soft penalty function of Eq. (5.31) in Section 5.4.3 can be used. As shown in Figure 7.6, we let A_k and $MaxA_k$ be total area of property piece k and its maximum allowable area affected by the alignment, respectively. $MaxA_k$ is initially set to be A_k for the property piece k inside the *HFB*, and be 0 for the property outside it; $MaxA_k$ can be interactively manipulated by the model users with the developed IDPM.

Note that similar forms of the soft penalty function, Eq. (5.31) are also used in the prescreening and repairing (P&R) method in Chapter 8 to control the solution alignments which are insufficient to accommodate required curve length. This penalty function is intended to smoothly guide the search in the optimization model. A penalty is assigned to the objective function value of the alignment if it violates the constraints.

Table 7.2 presents spatial attributes of an example study area map, and they are created from the IDPM interactively with a model user. Rows shaded in the table represent property pieces in the defined horizontal feasible bound (*HFB*). As stated previously, each property piece k has index variables, U_k and E_k, identifying whether it is inside U and E, respectively. There are unit property cost and area of k (denoted by C_k and A_k, respectively) in the table to calculate the right-of-way cost of the alignment. In addition, $MaxA_k$ and *Land-use* are also included in the attribute table to

Table 7.2 Attribute table of the study area map created from IDPM

Shape	k	U_k	E_k	C_k	A_k	$MaxA_k$	Land use
Polygon	1	1	0	0.15	3, 504	3, 504	Farm
Polygon	2	0	1	0.01	1, 000	0	Wetland
Polygon	3	1	0	10.20	2, 035	200	Resident
Polygon	4	1	0	11.04	890	100	Resident
Polygon	5	0	0	0.25	4, 082	0	Park
Polygon	6	1	1	0.12	1, 730	0	Cemetery
Polygon	7	0	0	13.44	2, 150	0	Commercial
Polygon	8	0	0	12.63	1, 830	0	Resident
Polygon	9	1	0	0.02	1, 632	1, 632	Stream
Polygon	10	0	1	2.16	1, 024	0	Historic
Polygon	11	1	0	0.88	851	100	Historic
Polygon	12	0	1	0.10	3, 730	0	Cemetery
·	·	·	·	·	·	·	·
·	·	·	·	·	·	·	·
·	·	·	·	·	·	·	·

reflect the user-defined constraints and to estimate environmental impacts of the alignment, respectively.

The horizontal feasible gate (HFG) method discussed in this chapter can also be applied to the fixed points in which a new alignment intersects with an existing road and stream or user-specified points. Each of those may require different specific constraints. For instance, constraints might limit the number of intersections if an alignment should not intersect an existing highway more than twice. Constraints might also limit the minimum vertical clearance if the alignment should pass over the existing highway. The proposed approach is applicable to many other cases if corresponding GIS data are available.

7.3 Feasible Gates for Vertical Alignments

To represent the vertical feasible gates (VFG) of an alignment, we again employ the orthogonal cutting plane method, which is an extension of the vertical cutting line concept to the three-dimensional (3D) alignment optimization. We first let the *HZ* plane be a coordinate system designed to represent ground and road elevation along a horizontal alignment. The *H* and *Z* axes represent road distance and elevation along the horizontal alignment, respectively. We now define a vertical alignment on the *HZ* plane.

Let $Start_V = (H_0, Z_0)$ and $End_V = (H_{n+1}, Z_{n+1})$ be start and end points of the vertical alignment, respectively where $H_0 = 0$ and Z_0, H_{n+1}, and Z_{n+1} are assumed to be known. Then, the set of vertical points of intersection (denoted as VPI_i for) can be defined as $VPI_i = (H_i, Z_i)$ as shown in Figure 7.7. The set of the consecutive points generally outlines the track of the vertical alignment, while linking each pair of successive points with a straight line produces a piecewise linear trajectory of the alignment.

A set of vertical feasible gates for *VPI*'s, denoted by V_i for all $i = 1, \ldots n_{PI}$ are placed in the orthogonal cutting planes (denoted by OC_i for all $i = 1, \ldots n_{PI}$) and bounded by upper and lower bounds, Z_i^{LB} and Z_i^{UB} for all $i = 1, \ldots n_{PI}$, respectively. Those bounds are determined with the elevations at the previous and subsequent intersection points and a pre-specified maximum gradient, g_{max}. The Road Elevation Determination Procedure is summarized in Section 7.3.1. Notations used in this procedure can be found in Table 7.3.

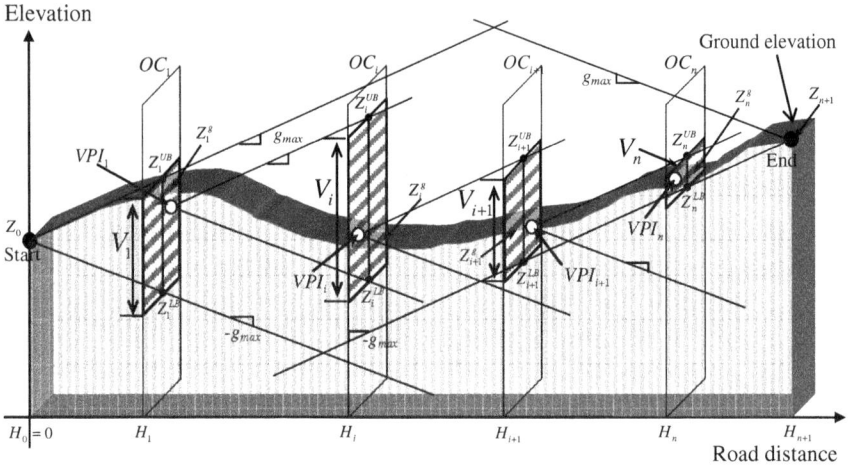

Figure 7.7 Representation of vertical feasible gates

7.3.1 Road Elevation Determination Procedure

Given Z_0, Z_{n+1}, g_{max}, and H_i, find Z_i for $i = 1, \ldots n_{PI}$

STEP 1: Calculate $tempL_i^1$ and $tempU_i^1$

$$tempL_i^1 = Z_{i-1} - (H_i - H_{i-1})\frac{g_{max}}{100}$$

$$tempU_i^1 = Z_{i-1} + (H_i - H_{i-1})\frac{g_{max}}{100}$$

STEP 2: Calculate $tempL_i^2$ and $tempU_i^2$

$$tempL_i^2 = Z_{n+1} - (H_{n+1} - H_{i-1})\frac{g_{max}}{100}$$

$$tempU_i^2 = Z_{n+1} + (H_{n+1} - H_{i-1})\frac{g_{max}}{100}$$

STEP 3: Calculate Z_i^{LB} and Z_i^{UB}, and then go to either STEP 4-1 or STEP 4-2

$$Z_i^{LB} = Max(tempL_i^1, tempL_i^2)$$

$$Z_i^{UB} = Min(tempU_i^1, tempU_i^2)$$

Table 7.3 Notations used in road elevation determination procedure

Notation	Descriptions
$VPI_i =$	i^{th} vertical point of intersection; $VPI_i = (H_i, Z_i)$, for $i = 1, \ldots n_{PI}$
$OC_i =$	i^{th} orthogonal cutting plane, for $i = 1, \ldots, n$
$V_i =$	vertical feasible gate where VPI_i is generated, for $i = 1, \ldots n_{PI}$
$H_i =$	H coordinate of $VPI_{,i}$, for $i = 1, \ldots n_{PI}$
$Z_i =$	Z coordinate at VPI_i, for $i = 1, \ldots n_{PI}$
$Z_i^g =$	ground elevation at H_i for $i = 1, \ldots n_{PI}$
$g_{max} =$	maximum gradient (%) defined by the model users
$Start_V =$	start point of a vertical alignment; $Start_V = (H_0, Z_0)$ where H_0 and Z_0 are given
$End_V =$	endpoint of a vertical alignment; $End_V = (H_{n+1}, Z_{n+1})$ where H_{n+1}, is alignment length and Z_{n+1} is given
$tempL_i^1 =$	provisional lower bound of Z_i based on Z_{i-1}
$tempL_i^2 =$	provisional lower bound of Z_i based on Z_{i+1}
$tempU_i^1 =$	provisional upper bound of Z_i based on Z_{i-1}
$tempU_i^2 =$	provisional upper bound of Z_i based on Z_{i+1}
$Z_i^{LB} =$	lower bound of Z_i for all $i = 1, \ldots n_{PI}$
$Z_i^{UB} =$	upper bound of Z_i for all $i = 1, \ldots n_{PI}$
$r_c[A, B] =$	a random value from a continuous uniform distribution whose domain is within the interval $[A, B]$

STEP 4-1: Find Z_i as close as Z_i^g

$$\text{Case 1}: \ Z_i = Z_i^{LB} \text{ if } Z_i^g < Z_i^{LB}$$

$$\text{Case 2}: \ Z_i = Z_i^g \text{ if } Z_i^{LB} \leq Z_i^g \leq Z_i^{UB}$$

$$\text{Case 3}: \ Z_i = Z_i^{UB} \text{ if } Z_i^g > Z_i^{LB}$$

STEP 4-2: Find Z_i randomly between Z_i^{LB} and Z_i^{UB}

$$Z_i = r_c[Z_i^{LB}, Z_i^{UB}]$$

Note that if the new alignment must pass through a certain point (e.g., a cross-point with an existing road), at which elevation is Z_{cp}, with a minimum vertical clearance (ΔH), its elevation at the point may be found from $Z_i = r_c[Z_i - \Delta H, Z_{cp} + \Delta H]$. Note that Eq. (5.39) can be used for estimating a penalty cost for violating the maximum gradient-constraint, and it can be rewritten as Eq. (7.17):

$$C_{PD^{VG}} = \sum_{i=1}^{n_{VC}} \left[\beta_{VG_0} + \beta_{VG_1} \left(\frac{Z_{i+1} - Z_i}{H_{i+1} - H_i} \times 100 - g_{max} \right)^{\beta_{VG_2}} \right] \quad (7.17)$$

where, $C_{PD^{VG}}$ = penalty associated with violating the maximum gradient constraint; n_{VC} = total number of vertical curve sections in the highway alignment generated; β_{VG_0}, β_{VG_1}, and β_{VG_2} are penalty parameters for the maximum-gradient constraint; Other terms are defined previously.

7.4   Example Study

The HAO model has been applied for a real-world road project, called Brookeville Bypass (on which Chapter 12 provides more information) to demonstrate the performance of the feasible gate (FG) method. Two example scenarios are tested in this model application; one is alignment search with the FG method, while the other does not employ the FG method. The baseline major design standards used in this example study are a two-lane road with a 40-foot cross-section (11 feet for lanes and 9 feet for shoulders), a 50 mph design speed, 5% maximum allowable gradient and 6% maximum superelevation. The model runs for 300 generations for each case. The user-specifiable maximum deflection angle, θ_{max} for calculating the allowable offset, D_{offset} is set at $\pi/2$ in this example.

To incorporate the horizontal feasible gate (HFG) method in the model, MDProperty View[1] is used as the base study area map and various land-use layers (such as wetlands, historic districts, and residential areas) and a horizontal feasible-boundary map defined by the model users are superimposed on the map. As shown in Figure 7.8, five horizontal feasible gates for PI's realistically represent the user-defined geographical boundary. The allowable offset, which is calculated based on the $\theta_{max}(=\pi/2)$ and the minimum curve radius (R_{H_m}) for the 50 mph design speed, is added to every feasible gate. Example solution alignments generated with the HFG method are successfully placed within the defined horizontal feasible bound. Note

---

[1] MDProperty View, developed by Maryland Department of Planning, is "a visually accessible database that allows people to interact with a jurisdiction's property map and parcel information using GIS software." (www.mdp.state.md.us/data)

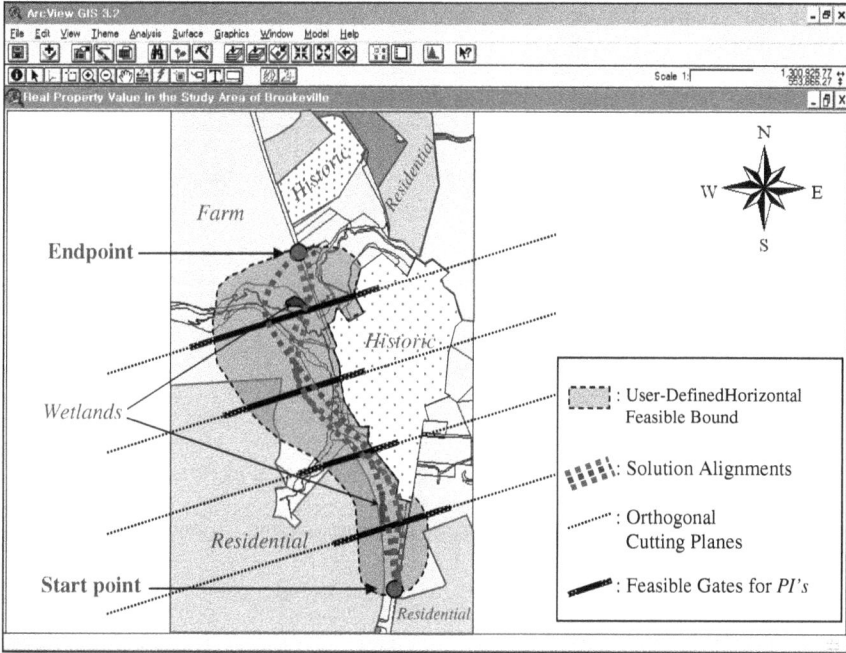

Figure 7.8 Example solution alignments obtained with FG methods for the brookeville project

that the 5% maximum allowable gradient is used for determining the vertical feasible gate (VFG) at every VPI in this example.

To test how fast each method (with vs. FG method) finds a reasonable solution, we set a solution boundary based on the optimized solution obtained with 1,000 generations for the same example problem. A "reasonable solution" is defined to be within 2% of the best known solution. Table 7.4 shows that the model tested without the FG method finds a reasonable solution in 5,311 seconds (88.52 minutes). However, with the FG method the model finds such a solution in 3,831 seconds (63.85 minutes), with 27.87% savings in computation time. It is noted here that such a computation time saving can significantly be improved if the scale of the road project is enlarged (e.g., the airline distance between endpoints is longer and geographic entities comprised in the study area increase). In this Brookeville example case, the size of the horizontal study area and the airline distance between two endpoints shown in Figure 7.8 are 3,600 feet $\times$ 8,400 feet

Table 7.4 Computation time comparison with/without FG methods for Brookeville Bypass project

Case	Total cost**	
	With FG method	Without FG method
Total cost of the solution alignment which first enters the 2% bound of the best-known solution ($)*	$4,387,209	$4,387,534
(% of the best solution)	102.00%	102.00%
Computation time to reach the 2% bound of the best-known solution (sec)	3,831 sec	5,311 sec
Computation time (%)	72.13%	102.00%

*The optimized solution obtained after 1,000 generation (total cost = $4, 301, 307) is assumed to be the best alignment for the Brookeville Bypass project.
**Total cost includes only agency cost (C_{T_Agency}) components for the Brookeville Bypass project.

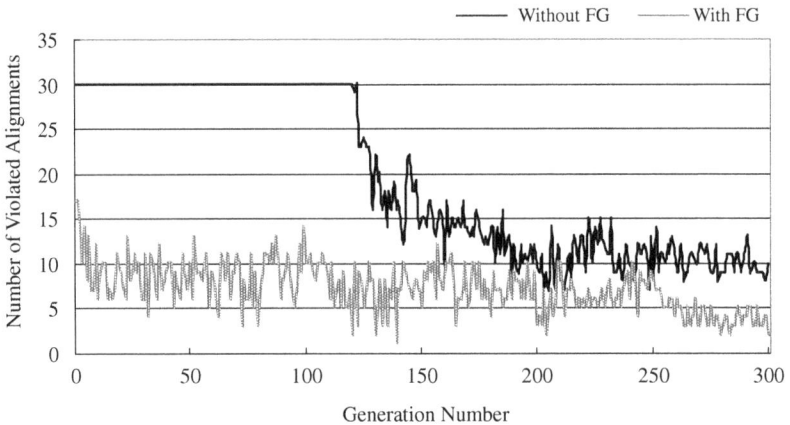

Figure 7.9 Number of solution alignments violating user-defined constraints over successive generations

Note: 30 solution alignments are generated in every generation

$(1, 097$ meters $\times 2, 560$ meters) and $4,003$ feet $(1,220$ meters), respectively. The study area comprises about 650 geographic entities, including private properties and roads.

Figure 7.9 shows how the solution quality improves over successive generations. With the FG method, the numbers of solution alignments

Table 7.5 Solution quality comparison with/without the FG methods for Brookeville Bypass project

Original

Generation #	Total cost ($)	Length-dependent*	Right-Of-Way	Total cost breakdown ($)						Road length (ft)
				Earthwork	Bridge	Grade-separation	C_{PE}**	C_{PDV}***	C_{PDH}****	
25	1,171,215,706	1,667,371	23,145,300	32,605,080	958,640	30,201	1,043,540,000	69,034,532	234,582	4168
50	344,967,932	1,850,606	55,259	4,984,417	892,925	109,864	235,870,600	101,087,400	116,861	4627
100	342,174,338	1,776,829	52,208	2,312,828	915,010	88,891	235,868,500	101,086,500	73,572	4442
150	6,563,385	1,736,228	49,648	3,678,298	916,665	89,449	48,903	12,048	32,146	4341
200	4,602,704	1,735,343	49,764	1,787,728	922,045	76,993	14,040	7,539	9,252	4338
250	4,435,934	1,734,040	49,851	1,660,253	911,240	76,993	3,557	0	0	4335
300	4,358,150	1,734,920	49,747	1,600,142	906,021	67,320	0	0	0	4337

with FG

Generation #	Total cost ($)	Length-dependent*	Right-Of-Way	Total cost breakdown ($)						Road length (ft)
				Earthwork	Bridge	Grade-separation	C_{PE}**	C_{PDV}***	C_{PDH}****	
25	6,039,951	1,792,966	52,309	3,011,272	925,805	83,778	10,000	0	163,821	4482
50	4,978,707	1,778,711	51,797	2,015,133	896,290	78,589	11,873	0	146,314	4447
100	4,643,786	1,730,160	49,442	1,872,422	894,305	55,358	20,382	0	21,717	4325
150	4,394,354	1,724,800	49,462	1,612,529	894,470	64,157	5,648	0	43,289	4312
200	4,344,982	1,720,800	49,816	1,599,630	897,162	67,253	0	0	10,321	4302
250	4,328,432	1,720,720	49,459	1,599,586	894,510	64,157	0	0	0	4302
300	4,328,432	1,720,720	49,459	1,599,586	894,510	64,157	0	0	0	4302

*The length-dependent cost, proportional to alignment length (e.g., pavement and road superstructure costs)
**Penalty associated with environmentally and geographically sensitive areas taken by the alignment
***Penalty cost for violating design constraints of vertical alignments
****Penalty cost for violating design constraints of horizontal alignments
Note: The HAO model runs on Pentium 4 CPU 3.2 GHz with 2GB RAM

violating the specified constraints, which include the user-defined horizontal and vertical bound constraints (i.e., geographically untouchable and partially untouchable areas and maximum gradient constraints), significantly decrease in early generations as shown in the figure. About 25% of the solutions with the FG method violate those constraints; however, most solutions obtained without the FG method have the constraint violations in early generations.

Such an effect can also be found in Table 7.5, which shows total cost breakdowns for the solution alignments at intermediate generations. The solution improvements (i.e., total cost improvements including various cost components) with the proposed FG method level off earlier than with the original method. Reasonably good solutions (defined to be within 2% bound of the best known solution) were found between 150 and 200 generations with the FG method, rather than 250 to 300 generations obtained without the FG method. This can be interpreted to indicate that the search process in the model now avoids the severely infeasible solutions much sooner and concentrates on refining good solutions with the FG method. C_{PE} (i.e., the penalty cost for violating environmentally and geographically sensitive areas) guides horizontally feasible alignments more effectively with the FG method. Such a penalty cost slightly affects the total costs of the solution alignments in an early stage of the search process since the solutions slightly exceed the specified allowable limit of areas. However, the penalty soon disappears in later generations. In addition, C_{PD^V}, which indicates a penalty cost for violating the bound constraints that guide vertically feasible alignments, does not influence the total cost through the entire generations since the FG method guides the model to avoid producing vertical alignments outside the feasible gates.

7.5 Summary

An efficient optimization method called feasible gate (FG) method has been developed to improve the computation efficiency and solution quality of the alignment optimization process. It improves the search efficiency of the model by restricting the model's search space (horizontally and vertically) so as to maximize the chance that alignments satisfying certain environmental, user preferences and geometric constraints are generated.

This is achieved by generating points of intersection (PI's) for alignments only within some appropriately limited subsets ("gates") of the orthogonal cutting planes. A customized GIS module (IDPM) is also developed for integrating the proposed method and the HAO model.

Two test examples with a real road project show how the proposed method improves the model's solution quality and reduces its computation time. Through a realistic application of the model with the FG method, it is found that the model's computation time and solution quality are improved throughout the search process. It is noted that the improvement due to the FG method can significantly increase if the scale of the road project is enlarged (e.g., if the number of geographic entities in the study area increases).

The HAO model can represent a complex road project more realistically and evaluate numerous alignments that satisfy various user preferences since the FG method assists the model in narrowing its horizontal and vertical feasible bounds based on the specified conditions including user preferences. Thus, it can focus sooner on refining the feasible alignments and provides the optimized solutions much faster. The FG approach is expected to be especially applicable in improving existing roads, such as by widening them within very limited bounds, besides optimizing completely new alignments. Some caution is required in using the FG method. The effect of the FG method would be negligible if the allowable offset (D_{offset}) added to the horizontal gates is excessive; i.e., the horizontal feasible gates for PI's might cover the entire search space of original method if the offset is excessive. On the other hand, it is possible to lose good candidate alignments if the offset is too short. Excellent solutions may run near borders between the specified feasible bounds and others, as shown in Figure 7.5(b).

Chapter 8

Prescreening and Repairing in Highway Alignment Optimization

Another efficient solution search algorithm (prescreening and repairing (P&R)), which is also developed for enhancing the computation efficiency and solution quality of the highway alignment optimization (HAO) model, is introduced in this chapter. The key idea of this method is to repair (before the detailed evaluation) any candidate alignment whose violations of applicable constraints (here mainly design constraints) can be fixed with reasonable modifications, but discard that alignment (by using a penalty method) and avoid the detailed evaluation procedure if its violations of constraints are too severe to repair. This chapter starts with Section 8.1, a research motivation of the P&R method learned from the HAO model application to a real highway project. In Section 8.2 the methodology of the P&R method is described. Two examples of the HAO model application to an actual highway project (called Brookeville Bypass) are tested in Section 8.3 to demonstrate the capability of the method. Finally, Section 8.4 summarizes the results.

8.1 Research Motivation of Prescreening and Repairing

Table 8.1 shows computation time breakdown for an optimization result obtained from the GA-based HAO model application to the Brookeville Bypass project in Chapter 8. In this application, the user cost components (including travel time, vehicle operating, and accident costs) are suppressed from the model objective function to ensure comparability with the normal evaluation criteria used by the Maryland State Highway Administration (MDSHA). Thus, five major supplier costs (i.e., right-of-way, construction and pavement, earthwork, grade separation, and bridge costs) and the

Table 8.1 Computation breakdown for an optimized result from the Brookville Bypass example

	Computation breakdown
Total alignment generation time	0.85%
Total alignment evaluation time	99.15%
Total program running time	100.00%
Total evaluation	
Right-of-way/environmental impacts via GIS	99.98%
Length-dependent*	0.00%
Earthwork	0.02%
Grade separation (for crossing existing roads)	0.00%
Bridge (for crossing rivers)	0.00%
Penalty	0.00%
Subtotal	100.00%

*The length-dependent cost represents cost proportional to alignment length, including pavement cost and road-superstructure cost

alignment's impacts on environmentally sensitive areas are considered in the solution evaluation procedure.

As shown in Table 8.1, most of the model computation time is spent on the detailed evaluation procedure; the total evaluation takes 99.15% of the total program running time, while only 0.85% of the time is taken by alignments generation. More precisely, the table shows that most (99.98%) of the evaluation time is devoted to the right-of-way cost and environmental impacts estimation, which are computed by a customized GIS module incorporated in the model. The evaluation time for the other cost components (0.02% of the total evaluation time) is almost negligible. It is noted that such a high evaluation time is mainly caused by the GIS process where many infeasible solutions generated by the GAs are evaluated. If the infeasible solutions can be prescreened before the GIS evaluation so that only feasible ones are evaluated, the computation time is significantly improved.

Figure 8.1 illustrates how many infeasible solutions are generated in the GAs-based highway alignment optimization (HAO) for the Brookeville application. They violate some model constraints (i.e., highway design standards associated with design speeds, such as minimum horizontal curve radius, minimum length of vertical curves, and sight distance). Note that with different design standards associated with given design speeds (e.g.,

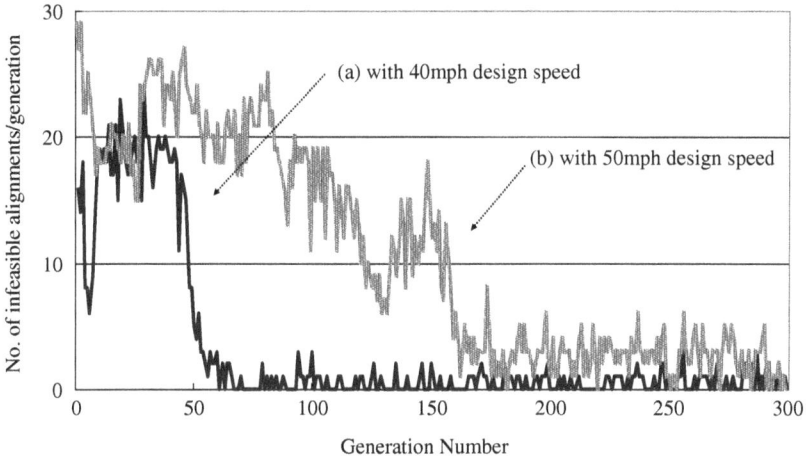

Figure 8.1 Number of solution alignments violating design constraints over successive generations for the Brookeville Bypass example

40 and 50 mph), the model searched for solution alignments over 300 generations for each case, by producing 30 solutions every generation. As shown in Figure 8.1, many solution alignments violate the design standards in early generations, but the fraction of the violated solutions tends to decrease over successive generations. In addition, the fraction is higher and more persistent when higher (i.e., more restrictive) design standards are used. This indicates that many candidate solutions generated by the GAs-based HAO model are not feasible, and they are more likely to violate the constraints if the applicable design standards are more constraining.

The HAO model relied on a penalty approach to guide the search toward better solutions. Thus, in the detailed evaluation procedure it assigns penalties to the cost components if the solution alignments generated from the GAs violate the corresponding constraints, and eventually screens out the candidate solutions whose constraint violations are significant. However, applying the detailed evaluation procedure (including the GIS computation) to all generated solutions (both feasible and infeasible) is computationally inefficient since obviously the infeasible solutions cannot be good candidates without any further modifications. That is why a direct constraint handling technique, called P&R method is developed and incorporated in the HAO model for efficiently controlling the geometric design constraints.

Using an artificial grid map instead of a real GIS database might be another possible way to speed up the model's computation time, and thus it was considered at early stages of the HAO model development. However, the GIS is very critical in the present model since it provides realistic right-of-way cost and environmental impact estimates for highway alternatives.

8.2 Prescreening and Repairing for Alignments Violating Design Constraints

As discussed in Chapter 4, 3D coordinates (xyz) of PI's are randomly created along the corresponding orthogonal cutting planes in the HAO model. Circular horizontal curves and parabolic vertical curves are then fitted. Horizontal transition curves can be added if desired. The curve fitting process, originally developed from Jong (1998), follows immediately after series of PI's and resulting tangents between those points are obtained. Ideally, a tangent section must be long enough to contain the required curve lengths which are determined with a design speed and deflection angle at PI's. However, in the original curve fitting process, the curve lengths at both ends are reduced to preserve a continuous alignment if a tangent is insufficient to accommodate the required curve lengths. The resulting alignments from the curve fitting process may not satisfy design constraints, and thus the model unnecessarily performs the time-consuming evaluation process for the infeasible solutions with penalties.

8.2.1 P&R Basic Concept

The prescreening and repairing (P&R) algorithm is developed to avoid such computational inefficiency, by allowing the detailed evaluation procedure only for the feasible solutions. The P&R method is intended for finding the segments that do not satisfy design constraints along the solution alignment, and fix that alignment by shifting the location of the corresponding PI's of the violated segments before undertaking any detailed evaluation procedure, but skip that evaluation procedure completely if the violations are too severe to repair. Thus, the model can concentrate more on refining good solutions. Figure 8.2 illustrates the concept of the P&R method.

The overall degree of design constraint violation in a candidate alignment can be represented by the percentage of the violated (i.e., infeasible)

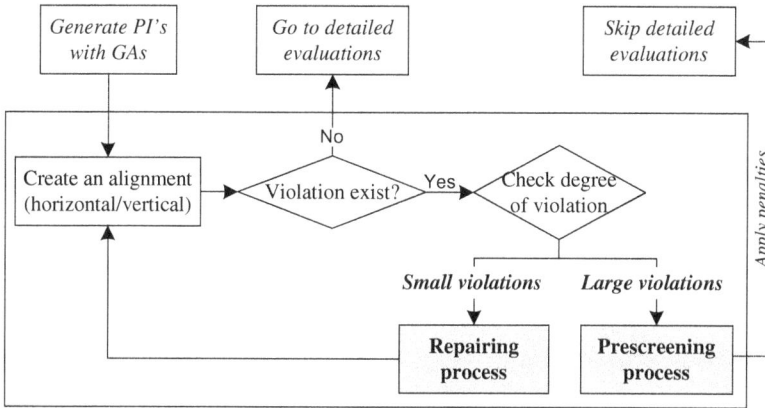

Figure 8.2 Prescreening and repairing concept

curve segments or by that of their corresponding PI's among the total PI's generated by the model. We introduce a user-specified parameter, denoted as F_i, to determine whether to repair an infeasible alignment. In the example study of the next section, we use $F_i = 50\%$ as a threshold value for distinguishing large violations from small ones. For instance, if five horizontal PI's are designed to create the solution alignments of the model, those alignments with three or more PI's corresponding to violated segments are classified as large violations and screened out from the detailed evaluation procedure. With the P&R method, the model repeats the repairing process until the alignments are fixed; however, if a violation is large, the alignment is discarded and the detailed evaluation procedure is skipped. Figure 8.3 shows a flow chart of the P&R method for the horizontal alignment evaluation.

8.2.2 Determination of Design Constraint Violations

Let $\mathbf{PI_i} = (x_i, y_i)$, be i^{th} horizontal point of intersection and Df_i^h be horizontal tangent deficiency between $\mathbf{PI_i}$ and $\mathbf{PI_{i+1}}$. It is noted that Df_i^h is used as a measure to calculate significance of the curve fitting violation and calculated as:

$$Df_i^h = (L_{PC_i} + L_{PC_{i+1}}) - \|\mathbf{PI}_{i+1} - \mathbf{PI}_i\|, \text{ If only a circular curve exists}$$

$$(8.1)$$

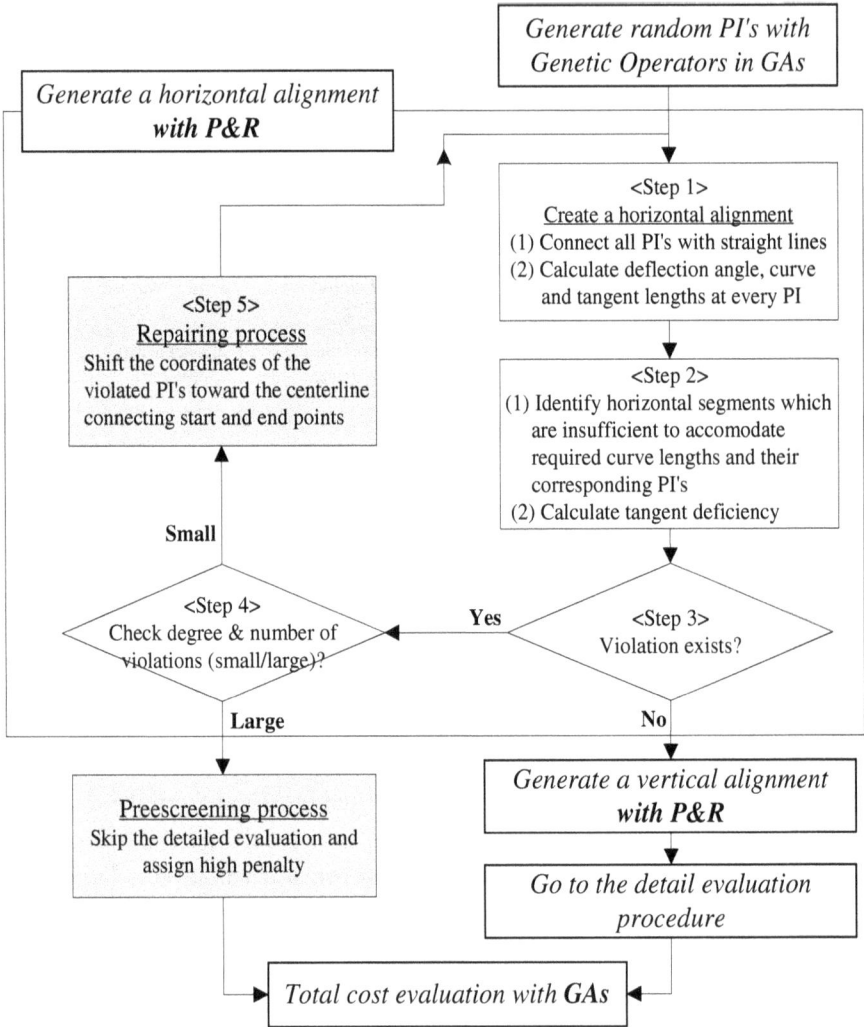

Figure 8.3 Prescreening and repairing process for horizontal alignment generation and evaluation

$$Df_i^h = (L_{ST_i} + L_{TS_{i+1}}) - \|\mathbf{PI}_{i+1} - \mathbf{PI}_i\|, \text{ If only a circular curve exists}$$
$$(8.2)$$

where, Df_i^h = horizontal tangent deficiency between $\mathbf{PI_i}$ and $\mathbf{PI_{i+1}}$

L_{PC_i} = tangent distance from point of circular curve ($\mathbf{PC}_i$) to $\mathbf{PI}_i$

L_{ST_i} = tangent distance from $\mathbf{PI}_i$ to the endpoint of transition curve ($\mathbf{ST}_i$)

L_{TS_i} = tangent distance from the beginning of transition curve ($\mathbf{TS}_i$) to $\mathbf{PI}_i$

$\| \ \|$ = norm (or length) of a vector

Other terms are defined previously.

Note that the following simple equation can be used to compute the tangent distance (L_{PC_i}) when there are no transition curves in a horizontal curved section of the alignment:

$$L_{PC_i} = R_{H_i} \tan \left(\frac{\theta_{PI_i}}{2} \right) \qquad (8.3)$$

$$\theta_{PI_i} = \cos^{-1} \left(\frac{(\mathbf{PI}_{i+1} - \mathbf{PI}_i) \cdot (\mathbf{PI}_i - \mathbf{PI}_{i-1})}{\|(\mathbf{PI}_{i+1} - \mathbf{PI}_i)\| \|(\mathbf{PI}_i - \mathbf{PI}_{i-1})\|} \right) \qquad (8.4)$$

where, R_{H_i} = horizontal curve radius at the ith horizontal curve section

θ_{PI_i} = deflection angle at $\mathbf{PI}_i$

$\cdot$ = Inner (dot) product used in vector operation

Other terms are defined previously.

Now, suppose that a violated horizontal curve segment, where the tangent deficiency exceeds zero (i.e., $Df_i^h > 0$), is identified on a solution alignment and its degree of design constraint violation is not too severe (i.e., percentage of the violated curve segments $\leq F_i$) as shown in Figure 8.4(a). Then, the violated alignment can be easily repaired by adjusting either $\mathbf{PI}_i$ or $\mathbf{PI}_{i+1}$ so that the tangent deficiency, Df_i^h is nullified or gets negative value. The PI that should be shifted between $\mathbf{PI}_i$ and $\mathbf{PI}_{i+1}$ is determined based on the magnitudes of deflection angles at those PI's. For instance, to repair the violated horizontal alignment shown in Figure 8.4(a), the repairing algorithm keeps adjusting $\mathbf{PI}_i$ if the deflection angle at $\mathbf{PI}_i$ is greater than that at $\mathbf{PI}_{i+1}$:

$$\text{If } \theta_{PI_i} \geq \theta_{PI_{i+1}}, \text{ then } \rightarrow \text{ Repair } \mathbf{PI}_i$$

$$\text{Otherwise } \rightarrow \text{ Repair } \mathbf{PI}_{i+1}$$

It should be noted that $\mathbf{PI}_{i+1}$ can also be selected to be adjusted during the repairing iteration any time when its deflection angle is greater than $\mathbf{PI}_i$'s. The iteration terminates if the violated segment is completely repaired. The

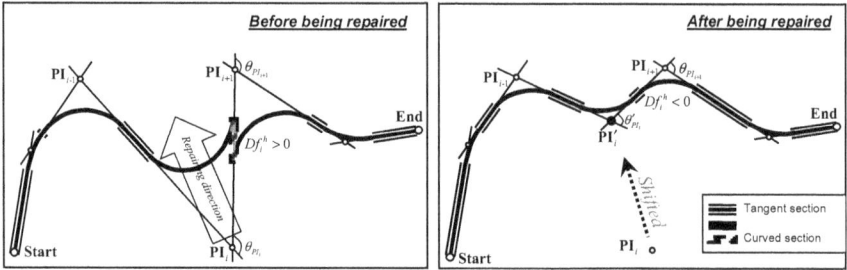

(a) Horizontal alignments before and after being repaired

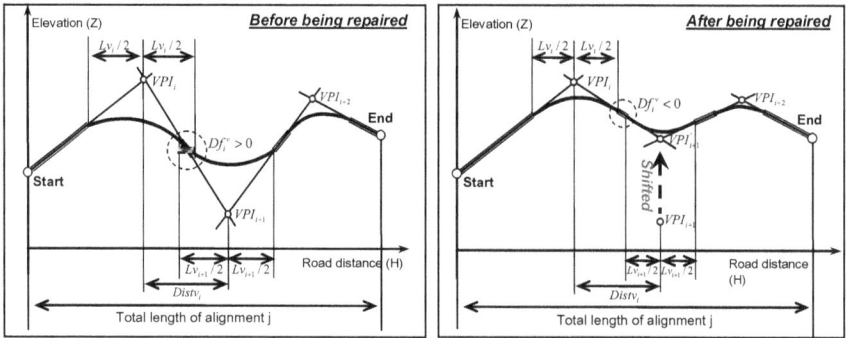

(b) Vertical alignments before and after being repaired

Figure 8.4 Solutions alignments before and after the repairing process: (a) for horizontal alignments and (b) for vertical alignments

P&R procedure is almost identically applied for the vertical alignments. Figure 8.4(b) shows the situation before and after the repairing process for vertical alignments. With the simple adjustment of the PI's and VPI's locations, we now can effectively fix the design violations in the solution alignments. It is noted that the locations of the adjusted PI's and VPI's are also random since their previous positions, which are originally obtained from the genetic operators, are randomly distributed along the orthogonal cutting planes.

Let VPI_i be i^{th} vertical point of intersection, denoted as $VPI_i = (H_i, Z_i)$, and Df_i^v be vertical curve-length deficiency at between VPI_i and VPI_{i+1}. Then, Df_i^v can be calculated as:

$$Df_i^v = \frac{1}{2}(L_{V_i} + L_{V_{i+1}}) - D_{VPI_i} \qquad (8.5)$$

where, L_{V_i} = vertical curve length at the ith vertical curve section

D_{VPI_i} = distance between VPI_i and VPI_{i+1}

In Eq. (8.5), the vertical curve length (L_{V_i}) should satisfy (i.e., be greater than or equal to) the minimum vertical curve length (L_{V_m}), calculated based on the stopping sight distance required on crest or sag vertical curve; i.e., $L_{V_i} \geq L_{V_m}$. The equation used for computing L_{V_m} can be found in AASHTO (2011). D_{VPI_i} can be measured with horizontal distance between VPI_i and VPI_{i+1} on HZ plane. Note that if an alignment is identified as highly infeasible or if a slightly infeasible alignment cannot be sufficiently repaired (or its violation worsens) during the repairing process, soft penalty functions, Eqs. (5.35) and (5.37) discussed in Chapter 5 can be used. These penalty functions can be rewritten with Df_i^h and Df_i^v respectively.

$$C_{PD^{HS}} = \sum_{i=1}^{n_{HC}} \left[\beta_{HS^0} + \beta_{HS^1} Df_i^{h \beta_{HS^2}} \right] \qquad (8.6)$$

$$C_{PD^{VL}} = \sum_{i=1}^{n_{VC}} \left[\beta_{VL^0} + \beta_{VL^1} Df_i^{v \beta_{VL^2}} \right] \qquad (8.7)$$

where, $C_{PD^{HS}}$ = penalty cost for violating design constraints of horizontal alignments

$C_{PD^{VL}}$ = penalty cost for violating the minimum length of vertical curve

n_{HC} = total number of horizontal curve sections of the highway alignment generated

n_{VC} = total number of vertical curve sections of the highway alignment generated

β_{HS^0}, β_{HS^1}, β_{HS^2} = coefficients used in computing $C_{PD^{HS}}$

β_{VL^0}, β_{VL^1}, β_{VL^2} = coefficients used in computing $C_{PD^{VL}}$

8.3 Example Study

Two scenarios are tested for the HAO model application to the Brookeville Bypass project to demonstrate the performance of the proposed P&R method. One is (i) the model application with the conventional penalty method for handling the design constraint violations, and the other is (ii) that with the P&R method. The baseline major design standards used in this case study specify a two-lane road with a 40 foot cross-section (12 feet for lanes and 8 feet for shoulders), a 50 mph design speed, 5% maximum allowable gradient and 6% maximum superelevation. MdProperty View 2004 is used as the baseline study area map and various land use layers (such as wetlands, historic districts, residential areas, and farms) are superimposed on the map.

In the presented Brookeville Bypass example case, the size of the horizontal study area and the airline distance between two endpoints are 3, 600 feet × 8, 400 feet (1, 097 meters × 2, 560 meters) and 4, 003 feet (1, 220 meters), respectively (see Figure 8.5). About 650 geographic entities (land properties, structures, roads, etc.) are comprised in the study area. Optimized alignments obtained with different number of PI's (4 to 7 PI's) are also presented in Figure 8.5. Note that an extensive sensitivity analysis

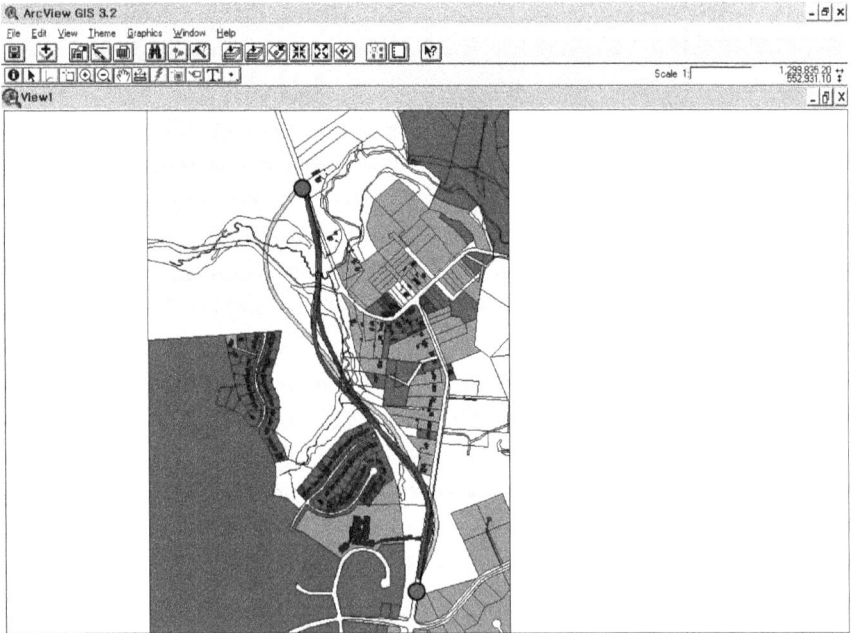

Figure 8.5 Study area and optimized alignments obtained from the model application to the Brookeville Bypass project

Source: Kang *et al.*, 2012.

to factors affecting geometry and location of the optimized alignments can be found in Chapter 12.

It should be noted here that we do not know the exact optimal solution to the given problem. However, since the goodness of the solutions obtained from the HAO model (which originally use the penalty method for handling design standard violations) has been statistically tested and proved its significance with many case studies (Jong *et al.*, 2000; Jong and Schonfeld; 2003), we assume that the optimized solution found from the model within 1,000 generations is the best known solution for the given problem. In addition, we set a solution boundary based on the best known solution for testing how fast the model incorporated with the P&R method finds a reasonable solution. Note that a "reasonable solution" is defined to be within the 2% bound of the best known solution.

Table 8.2 shows that the model tested with the original penalty method finds a reasonable solution in 5,311 seconds (88.52 minutes) while the model with the P&R method finds the solution in 4,077 seconds (67.95 minutes), with 23.23% computation time savings. This occurs mainly because the model with the P&R method skips the detailed evaluation procedure for solution alignments that violate the design constraints. It prohibits the model from exploring the GIS. Furthermore, the P&R method allows the model to consider more alignments than the original case for the same given

Table 8.2 Solution comparisons with and without P&R methods for the Brookeville Bypass project

		Original	With P&R
Program running time to reach the 2% bound of the best-known solution (sec)		5,311	4,077
Computation time saving (%)		(0.00%)	(23.23%)
Total alignments generated	No. of alignments that go to the detailed evaluation procedure	8,783	8,103
	(%)	(100.00%)	(54.32%)
	No. of alignments that skip the detailed evaluation procedure	0	6,814
	(%)	(0.00%)	(45.68%)
	Subtotal	8,783 $\longrightarrow$ 14,917	
	(%)	***69.84%***	
		(100.00%)	(100.00%)

number of generations (about 69.84% more). However, the original method evaluates all generated alignments including the infeasible ones; note that about 36.70% of the total generated alignments (i.e., $0.367 \times 8,783$) have design constraint violations.

Additionally, as shown in Figure 8.1, the fraction of the infeasible solutions would increase more if the design standards were stricter. The P&R approach would become increasingly advantageous as the design standards get higher (i.e., more constraining), since it prescreens and repairs an increasing fraction of the alignments.

We also test total cost improvements over the successive generations for each scenario. For both the cases, most of the improvements are found in the early generations. As presented in Figure 8.6, showing the convergence of both applications, there are no significant improvements in the total cost (less than 0.5% improvements) after about 220 generations. However, it is noted that the improvements with the proposed P&R method level off significantly earlier than that with the original penalty method. This indicates that after incorporation of the P&R method the model stays away from the severely infeasible solutions much sooner and concentrates on refining good solutions.

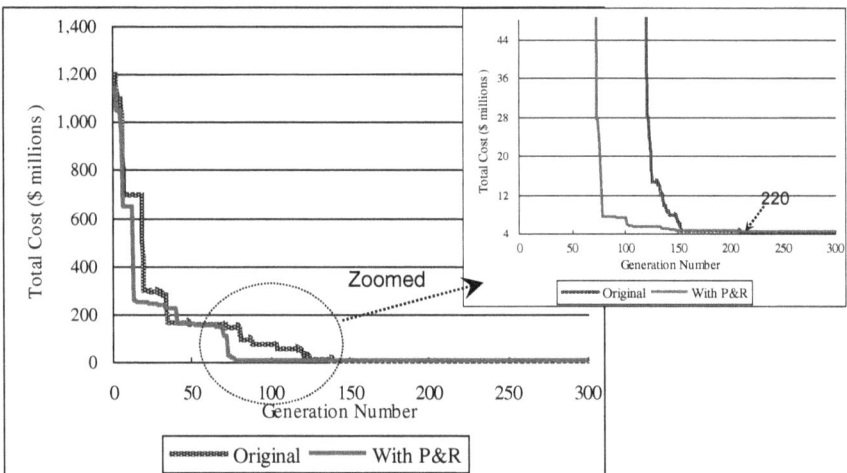

Figure 8.6 Changes in total cost over successive generations for different scenarios

8.4 Summary

An efficient constraint handling technique, combining prescreening and repairing (P&R), is developed to enhance the computation efficiency and solution quality of the GAs-based highway alignment optimization process. The key idea of the proposed P&R method is to repair (before the very detailed alignment evaluation) any candidate alignments whose violations of design constraints can be fixed with reasonable modifications. However, those infeasible alignments are discarded (prescreened) by avoiding the detailed evaluation procedure if the violations of constraints are too severe to repair. The proposed method is simple, but improves the model's computation time and solution quality significantly. Such improvements are demonstrated with a test example for a real highway project. Through the model's application to a real project, it has been shown that the model incorporating the P&R method can find a reasonably good solution alignment (within the 2% bound of the best known solution) much faster than the same model with the original penalty method, with about 23% computation time saving. In addition, the P&R method allows the model to consider about 70% more solutions than that with the penalty method for the same given number of generations. Note that the P&R approach becomes increasingly advantageous as the design standards get stricter since it prescreens and repairs an increasing fraction of the infeasible alignments.

The HAO model now has a two stage solution search process: (i) the prescreening and repairing and (ii) detailed evaluation stages. Further research might also consider a multi-stage screening process based on the relative computation time of the various cost components to increase computation efficiency. Such improvements are important for extending the model beyond single highway alignments to more complex problems, such as network optimization problems. Note that if the P&R method is combined with the FG method discussed in Chapter 7, the computation time for solving the complex highway alignment optimization (HAO) problem would be significantly faster.

Part III

Optimizing Simple Highway Networks: An Extension of Genetic Algorithms-based Highway Alignment Optimization

In Part III we further develop the GA-based HAO model, discussed in PART II, to extend its capabilities to optimize highway alignments within an existing road network. For that, the highway location selection and alignment optimization problem is reformulated as a bi-level programming problem to reflect changes in traffic flow patterns over the network due to its improvement with various highway alignment alternatives. The upper-level of the bi-level problem is formulated as the alignment optimization problem, and (2) the lower-level problem is defined as an equilibrium traffic assignment problem.

Part III starts with Chapter 9 — an overview of discrete network design problem, which is similar to but clearly distinguished from the highway location selection and alignment optimization problem. The bi-level optimization concept, formulation, and model structure to deal with optimizing highway alignments within a small roadway network are discussed in Chapter 10. Finally, a bi-level HAO model application to hypothetical examples is demonstrated in Chapter 11.

Chapter 9

Overview of Discrete Network Design Problems

The highway location selection and alignment optimization problem defined in Chapter 1 is somewhat similar to the network design problem (NDP) which deals with the optimal decisions on the improvement of an existing highway network in response to a growing demand for travel (Gao *et al.*, 2005). Roughly, the NDP can be classified into two different forms: discrete and continuous versions. The discrete version of the problem, known as DNDP, finds optimal (new) highways added to an existing road network among a set of predefined possible new highways while its continuous version, known as CNDP, determines the optimal capacity expansion of existing highways in the network. In whichever form, the objective of the NDP is usually to minimize total system travel cost while accounting for the route choice behaviors of network users. Note that models developed for the CNDP are not reviewed in this monograph since they are only distantly related to our problem (which considers the addition of a new highway to an existing network).

9.1 Bi-level Discrete Network Design Problems

In many studies (Bruynooghe, 1972; Steenbrink, 1974; LeBlanc, 1975; Johnson *et al.*, 1978; Pearman, 1979; Magnanti and Wong, 1984; Xiong and Schneider, 1992; Yang and Yagar, 1994; Yang and Lam, 1996; Yang and Bell, 1998; Yin, 2000; Lo and Tung, 2001; Meng *et al.*, 2004; Chen and Yang, 2004; Gao *et al.*, 2005; Sharma and Mathew, 2007), the DNDP is usually expressed by a bi-level programming problem in which the upper-level problem represents decision-making process of a network designer (e.g., transportation authority), and the lower-level problem represents route choice behavior of the network users under the designer's decision. Note that the bi-level DNDP is recognized as one of the most challenging problems in transportation (Magnanti and Wong, 1984; Yang and Bell, 1998) due to its computational difficulties. The DNDP is proven to be a NP-complete (NP: nondeterministic polynomial time) problem by Johnson *et al.* (1978).

In the traditional bi-level programming model for the DNDP, it is assumed that the system designers can affect the network users' path-choosing behavior by adding new highways, but cannot control them (i.e., the users make their decision in a user optimal manner). In addition, the traffic demand in the network is assumed to be given and fixed; however, the model allows changes in traffic flow over the network from the improvement of the road network by adding a new highway. A typical formulation of the bi-level programming problem used for the DNDP (Yang and Yagar, 1994) is described as follows:

Upper Level Problem

$$\underset{\mathbf{u}}{Minimize} \quad F(\mathbf{u}, \mathbf{v}(\mathbf{u}))$$

$$subject \ to \quad G(\mathbf{u}, \mathbf{v}(\mathbf{u})) \leq 0$$

where $\mathbf{v}(\mathbf{u})$ is implicitly defined by

Lower Level Problem

$$\underset{\mathbf{x}}{Minimize} \quad f(\mathbf{u}, \mathbf{v})$$

$$subject \ to \quad \mathbf{g}(\mathbf{u}, \mathbf{v}) \leq 0$$

In the above formulation, F and $\mathbf{u}$ are the objective function and decision vector of upper-level decision makers (system designer) respectively, while G is the constraint set of the upper-level decision vector. f and $\mathbf{v}$ are the objective function and decision vector of lower-level decision makers (users traveling in the network) respectively, while $\mathbf{g}$ is the set of constraints of the lower-level decision vector. It is noted here that $\mathbf{v}(\mathbf{u})$ is implicitly defined by the lower-level problem (i.e., the upper-level objective function F cannot be computed until $\mathbf{v}(\mathbf{u})$ is determined in the lower-level problem).

9.1.1 Upper-level DNDP

In the bi-level DNDP, the upper-level problem, which represents the decision-making of the system designer, usually can be formulated as a total cost minimization problem based on the equilibrium traffic flow found in the lower-level problem. Many studies have attempted to solve the upper-level optimization problem in different ways, such as with a decomposition method (Steenbrink, 1974), a Branch and Bound method (LeBlanc, 1975; Poorzahedy and Turnquist, 1982), simulated annealing (SA) based methods (Friesz *et al.*, 1992), genetic algorithm (GA)-based methods (Xiong and Schneider 1992; Yin, 2000; Chen and Yang, 2004; Sharma and Mathew, 2007), and others (Pearman, 1979; Gao *et al.*, 2005, etc.). Among them, the GA-based approach is the most popular one because of its simplicity and ability to handle large problems.

9.1.2 Lower-level DNDP

The lower-level problem, which represents the user route choice behavior, can be solved with different types of traffic assignment methods. Choices can be made between static and dynamic assignments, and between deterministic or stochastic assignments. In previous studies on the DNDP, different assignment methods are used for solving the lower level problem. Most studies (LeBlanc, 1975; Friesz *et al.*, 1992; Gao *et al.*, 2005; Sharma and Mathew, 2007; etc.) use the Frank-Wolfe algorithm, which is a deterministic (and static) user equilibrium method, to solve the lower-level problem. In Xiong and Schneider (1992), a neural network approach is used to carry out a deterministic user equilibrium assignment. The stochastic user equilibrium assignment is used in Chen and Alfa (1991), Davis (1994), Lo and Tung (2001), and Meng *et al.* (2004). Here, we adopt the Frank-Wolfe algorithm to obtain the equilibrium traffic flow pattern.

9.2 Comparison of Highway Alignment Optimization and Discrete Network Design Problems

In the relevant literature, there are two types of optimization problems which deal with highway location selection and alignment optimization for improving an existing roadway network. These are known as 1) the highway alignment optimization (HAO) problem and 2) discrete network design problem (DNDP). The first is a microscopic highway alignment design problem whose objective is to find the "best" alignments for a new highway connecting specified points and sections on existing roadways subject to a variety of requirements and constraints. Various highway cost items (e.g., right-of-way cost and earthwork cost), factors (e.g., terrain profiles and land-use), and constraints (e.g., geometric design standards) associated with road construction are included in the problem. Many mathematical models have been developed for optimizing either horizontal or vertical alignments or fully three dimensional (3D) alignments of new highways (Trietsch, 1987a, 1987b; Jong *et al.*, 2000; Easa *et al.*, 2002; Cheng and Lee, 2006; Lee *et al.*, 2009; Kang *et al.*, 2012). They employ methods of Calculus of Variations, Dynamic Programming, Linear Programming, Enumeration, and Genetic Algorithm (GA) for highway route optimization, and many possible highway alignments are generated and evaluated with the methods. Nevertheless, these models take into account only costs associated with road construction, and therefore the most economic alignment is presumed to be the "best". Traffic impacts on the existing roadway network due to a new highway implementation are not included among the evaluation criteria. Furthermore, most relevant models, except for the GA-based HAO model introduced in this monograph, do not reflect environmental impacts of new highways in the evaluation process.

The DNDP is a macroscopic highway design problem which seeks an optimal link addition to an existing road network among a set of predefined possible links. Its objective is usually to minimize total user travel cost in the network, accounting for the route choice behaviors of the network users. Many mathematical models have been developed to solve the DNDP (e.g., Steenbrink, 1974; LeBlanc, 1975; Chen and Yang, 2004; Gao *et al.*, 2005). These models can deal with larger networks than those studied in the HAO problems, and are usually expressed as a bi-level programming problem in which the upper-level problem represents decision-making process of a network designer while the lower-level problem represents route choice behavior of the network users. In the DNDP model, a conceptual road network with sets of nodes and arcs is used to represent trip generators (e.g., traffic cities) and new and existing highways. The DNDP model assumes that a set of predefined new links is given as a model input and identifies whether it should be linked in the existing network with two binary

Table 9.1 Comparison of highway alignment optimization and network design problems

	Highway alignment optimization	Network design problem
Scope	— Microscopic highway route selection and design	— Macroscopic highway network planning
Objective	— Find actual highway alignments that minimize costs associated with road construction	— Find a network configuration that minimizes network travel cost (normally, travel time cost)
Input	— Highway design specification	— Conceptual road network with sets of nodes/links
	— Spatial information of the study area (e.g., terrain, land-use, property value)	— Travel demand (i.e., origin/destination trip matrix)
Output	— Profiles of optimized highway alignments	— Conceptual road networks with selected new links
	— Cost items required for road construction	— Network travel cost
Advantage	— Generate and evaluate highway alignments	— Reflect drivers' route choice behavior
	— Reflect actual construction costs in evaluation	— Can deal with larger networks
Disadvantage	— Cannot reflect traffic impacts of highway alternatives in their evaluation process	— Cannot consider actual highway cost and constraints associated with road construction
		— Cannot generate different highway alternatives

integer values (e.g., "1" = add and "0" = do not add). However, it should be noted that such a macro-level model may be impractical to use in an actual highway project which mainly focuses on the detailed design of horizontal and vertical profiles of the new highway and its land-use impacts. No DNDP model which deals with detailed highway construction cost (e.g., earthwork and right-of-way costs) as well as geographic and environmental concerns has been proposed in the literature. Despite such limitations, the main advantage of the DNDP model is that it takes into account the traffic impact of a new link addition to the existing road network in its evaluation process, which is not considered in the models developed for highway route optimization. Table 9.1 shows basic differences between the DNDP and the highway alignment optimization problem.

Note that the HAO problem itself is very complex requiring a time consuming search process. If a traffic assignment process is added to solve the HAO problem, its computational burden increases further. Therefore, development of efficient solution search algorithms is essential for handling this larger problem. Efforts in computation time reduction of the upper-level problem are covered in PART II of this monograph where two efficient solution search methods (Feasible Gates and P&R methods) are presented. The network version of the HAO model (i.e., bi-level HAO model) is further discussed in Chapter 10.

Chapter 10

Bi-level Highway Alignment Optimization within a Small Highway Network

A new highway addition to an existing road network may be considered to improve the performance of that road network. Such a supply action increases capacity of the road network, and highway users may thus save travel time and vehicle operation costs. However, finding the new highway alternative that best improves the existing road network is a very complex problem since many factors affect it. Changes in network traffic flow patterns from the new highway addition, various costs associated with highway construction, as well as design specifications, safety, environmental impacts, and political issues affect the project. In addition, profiles and impacts of the new highway may depend significantly on how and where it is connected to the existing road network. Many previous studies (e.g., Steenbrink, 1974; Trietsch, 1987a, 1987b; Jong *et al.*, 2000; Chen and Yang, 2004; Gao *et al.*, 2005; Lee *et al.*, 2009; Kang *et al.*, 2012) have proposed various mathematical methods for solving highway network design and route optimization problems. Although some methods perform well in certain aspects, all are limited in the factors that they consider. We find no previous model that jointly evaluates traffic and environmental impacts of the new highway as well as optimizes highway location, construction cost, and horizontal and vertical profiles. This study integrates all these factors in optimizing highway alignments.

Finding new highways that best improve an existing roadway system can be described as a leader-follower game in which the system designers (i.e., highway planners and designers) are leaders and the highway users (i.e., motorists) who can freely choose their paths are the followers. In this process the system designers can influence but not control the route choice behavior of highway users. The system designers try to find an economical path that minimizes the total construction cost, while considering geometric design and geographical constraints. However, the traffic flow is determined by user decisions which can be approximated by the user equilibrium principle. To realistically represent such characteristics in the highway route optimization process, this monograph introduces a bi-level optimization method. In the bi-level method, the upper-level problem is defined as the highway alignment optimization problem in which best highway alternatives are identified based on a specified objective function. The lower-level is a user equilibrium traffic assignment problem seeking to optimize the travel time for the highway users in the network. This bi-level approach is superior to a method which optimizes only highway construction costs; furthermore, it can provide a much wider scope of objectives regarding various user costs including travel time, vehicle operation, and accidents costs.

10.1 Bi-level HAO Concept

Highway alignments are generally optimized by minimizing the costs associated with road construction. The highway costs are typically agency-related costs which include earthwork cost, right-of-way cost, pavement cost, etc. All these cost components can be key criteria of highway construction and should be evaluated simultaneously. It should also be noted that new highway construction increases the capacity of a road network, and thus decreases travel time and vehicle operating cost of motorists in that network. Furthermore, traffic patterns in the revised network may significantly vary depending on the new highway's length, location, and connection points to the network. Such variation is due to the highway users' path-choice behavior that cannot be directly controlled by system designers (i.e., highway agencies).

In order to incorporate the users' behavior into the highway route optimization process, a bi-level programming structure is developed here (see Figure 10.1). The upper-level problem of the bi-level structure represents a decision-making process of system designers, in which possible highway

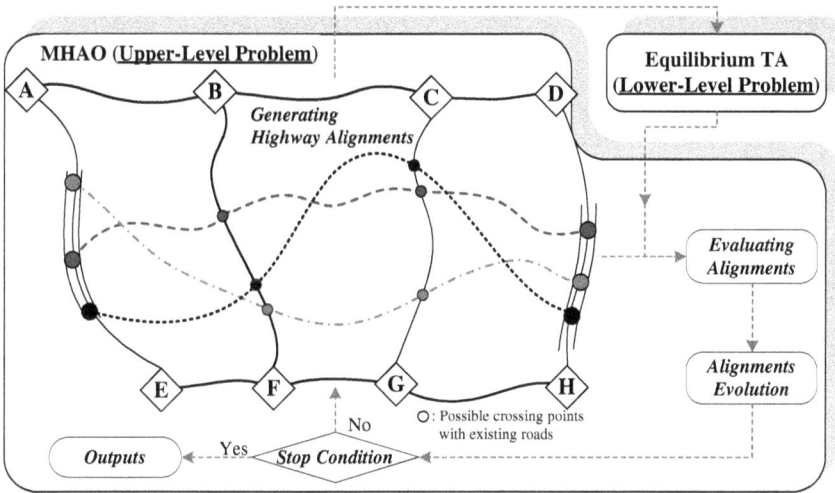

Figure 10.1   Bi-level highway alignment optimization concept

alternatives are generated and evaluated. The lower-level problem represents highway users' route choice behavior under the designers' decision (i.e., the alternative selected from the upper-level). Traffic impacts of the new highway on the existing network are estimated through a user equilibrium traffic assignment process (i.e., the lower-level problem), and the assignment results (e.g., equilibrium traffic flows and travel times) are used for computing the user cost, which is one of the evaluation criteria included in the upper-level problem. The concept of the bi-level highway route optimization is shown in Figure 10.1, and its generalized mathematical formulation can be expressed as in Eqs. (10.1) to (10.4).

Upper-Level (Objective: to find optimized highway alignments)

$$\text{Minimize } Z_{UL} = C_{T\_Agency} + C_{T\_User} + C_P \qquad (10.1)$$

Subject to: *highway design, geographical, and environmental constraints*

where Z_{UL} = upper level objective function value of the bi-level HAO; C_{T_Agency} = total agency cost required for a new highway development; C_{T_User} = total user cost; C_P = total penalty cost.

Lower-Level (Objective: to find user equilibrium flows of updated road networks)

$$\text{Minimize } Z_{LL} = \sum_a \int_0^{x_a} t_a(x_a) dx \quad \text{for user optimum} \qquad (10.2)$$

$$\text{Minimize } Z_{LL} = \sum_a x_a t_a(x_a) \quad \text{for system optimum} \qquad (10.3)$$

Subject to:

$$\begin{cases} \sum_k f_k^{rs} = q_{rs} & \text{for all } r, s \\ f_k^{rs} \geq 0 & \text{for all } k, r, s \\ x_a = \sum_r \sum_s \sum_k f_k^{rs} \delta_{a,k}^{rs} & \text{for all } a \in \mathbf{A} \end{cases} \qquad (10.4)$$

where Z_{LL} = lower level objective function value of the bi-level HAO; $\mathbf{A}$ = arc set of a given highway network, $a \in \mathbf{A}$; x_a = flow on arc a, $\mathbf{x} = (\cdots, x_a \cdots)$; t_a = travel time on arc a, $\mathbf{t} = (\cdots, t_a \cdots)$; f_k^{rs} = flow on path k connecting O/D pair r-s, $\mathbf{f}^{rs} = (\cdots, f_k^{rs}, \cdots)$; q_{rs} = trip rate between origin r and destination s, $r \in \mathbf{R}$, $s \in \mathbf{S}$; $\mathbf{N}$ = node set of a given highway network, $\mathbf{R} \subseteq \mathbf{N}$, $\mathbf{S} \subseteq \mathbf{N}$; $\delta_{a,k}^{rs}$ = indicator variable (1: if arc a is on path k between O/D pair r-s; 0: otherwise).

10.2 Upper Level of Bi-level HAO

Two types of decision variables are used in the upper level of the bi-level HAO model structure: (1) points of intersection (PI's) of new highway alignments and (2) distributed traffic flows on the network. The objective function of the upper-level problem primarily depends on these variables along with many other factors such as unit pavement cost, earthwork quantity, fuel cost, and land-use. Note that the decision variables, PI coordinates, are indirectly formulated in the upper-level objective function. To solve the upper-level problem, a genetic algorithm (GA) with customized genetic operators (Jong and Schonfeld, 2003) is employed in the model. The GA aims to generate the PI's of new highways, and ultimately finds optimized ones through an evaluation procedure based on the principles of natural evolution and survival of the fittest. The formulation of the upper-level alignment optimization problem includes an objective function and two constraints associated with new highway construction. The objective function (Z_{UL}) is defined as the sum of (1) the total agency cost, (2) the total user cost, and (3) the penalty and environmental costs as discussed in Chapter 5.

10.2.1 Highway Agency Cost for Bi-level HAO

The total agency cost consists of four major construction costs [length-dependent cost (C_L), right-of-way cost (C_R), earthwork cost (C_E), structure cost (C_S)] directly required at the initial stage of a new highway development and a maintenance cost (C_M) occurring throughout the life of the road alignment. All these cost components are important and sensitive to highway alignments, and should be simultaneously evaluated in the highway alignment optimization process. The mathematical formulations of the first five components (i.e., C_L, C_R, C_E, C_S, and C_M) are already defined in Chapter 5, and not duplicated in this chapter.

10.2.2 Highway User Cost for Bi-level HAO

Drivers' route choice behavior can change before and after a new highway addition to a highway network, and may vary for its different alternatives. To consider such variation in the alternative evaluation process, this monograph considers the user cost components of the different alternatives in the objective function of the bi-level HAO model, besides their agency costs presented in the previous section. Let $C^0_{T\_User}$ and $C^1_{T\_User}$ be total user costs before and after a new highway construction, respectively. Then, we may expect either positive or negative user cost savings (denoted as $\Delta C_{T\_User} = C^0_{T\_User} - C^1_{T\_User}$) from the highway. The highway users may save their travel time and fuel consumption by taking a new alternative route to their travel destinations; however, they may incur higher cost if an undesirable alternative is built. Three types of user cost saving that may be considered in the bi-level HAO model are presented in Table 10.1. However,

Table 10.1 Type of user cost savings in bi-level HAO

User cost saving components	Examples
Travel time cost saving	Decrease or increase in travel time
Vehicle operating cost saving	Decrease or increase in fuel consumption
Accident cost saving	Decrease or increase in number of accidents

accident cost saving is not further discussed in this chapter as it is too complex to quantify. However, it is important to note that the bi-level HAO model is designed with a modular structure in which various evaluation components can be easily added, suppressed, or replaced without changing the rest of the model structure. Thus, any available accident prediction relations or models can be incorporated in the model for estimating the accident frequency of new highways.

In the bi-level HAO model, a BPR function (AASHTO, 2003) is used for estimating link travel times between origin-destination (O/D) pairs. The base year O/D matrix is a given input, and the demand growth is also considered by multiplying the base year demand by a compound growth factor. It is assumed that the overall O/D flows in the highway network are stable with and without a new highway addition (i.e., $q_{rs}^0 = q_{rs}^1 \ \forall r, s (r \neq s)$ where q_{rs}^0 and q_{rs}^1 are trip rates between origin r and destination s before and after the new highway construction, respectively; $Q^0 = \sum q_{rs}^0$ and $Q^1 = \sum q_{rs}^1 \ \forall r, s (r \neq s)$); however, individual users within the road network can freely select their travel paths. A basic economic concept used for evaluating the user cost savings is presented in Figure 10.2. In it, superscripts 0 and 1 stand for conditions of a given highway network before and after a new highway addition, respectively. For example, F^{1j} represents a conceptual traffic performance function of the network after addition of highway alternative j. Given the assumption, mathematical relations between traffic and cost components are described in Eqs. (10.5) to (10.8).

$$C_{T\_User} = C_T + C_V \tag{10.5}$$

$$C_T = \sum_{a \in \mathbf{A}} (x_a t_a) [\mathbf{v} \cdot \mathbf{T} \cdot \mathbf{o}] \tag{10.6}$$

$$C_V = \sum_{a \in \mathbf{A}} (x_a L_a) [\mathbf{u}_a \cdot \mathbf{T}] \tag{10.7}$$

$$\mathbf{u}_a = \begin{bmatrix} p_{F_{Auto}} f_{a\_Auto} + m_{Auto} \\ p_{F_{Truck}} f_{a\_Truck} + m_{Truck} \end{bmatrix} \tag{10.8}$$

where C_T, C_V = travel time cost and vehicle operating cost, respectively; $\mathbf{v}$ = travel time values for auto and truck, $\mathbf{v} = [v_{Auto}, v_{Truck}]$; v_{Auto} and v_{Truck} are unit travel time value (\$/hr) of auto and truck drivers, respectively; $\mathbf{T}$ = traffic composition vector, $\mathbf{T} = [1 - T_{Truck}, T_{Truck}]$; T_{Truck} is the proportion of trucks in traffic flow stream over the network; $\mathbf{o}$ = average vehicle occupancy, $\mathbf{o} = [o_{Auto}, o_{Truck}]$; o_{Auto} and o_{Truck} are average vehicle occupancy for auto and truck drivers, respectively; $\mathbf{u}_a$ = unit vehicle operating cost for autos and trucks traveling on arc a; L_a = length of arc a in the highway network; $p_{F_{Auto}}$, $p_{F_{Truck}}$ = fuel prices ($/gallon) for auto and truck, respectively; f_{a_Auto}, f_{a_Truck} = fuel consumption of auto and truck, which can be estimated from their average speeds on arc a; m_{Auto}, m_{Truck} = maintenance cost of auto and truck, respectively.

Based on the concept used in Figure 10.2 and Eqs. (10.5) to (10.8), highway users' travel time and vehicle operating cost savings within a network with the addition of different highway alternatives are evaluated in the optimization process.

The cost evaluation procedure presented in each sub-section below modifies and adjusts the basic steps of the user benefit analysis proposed in

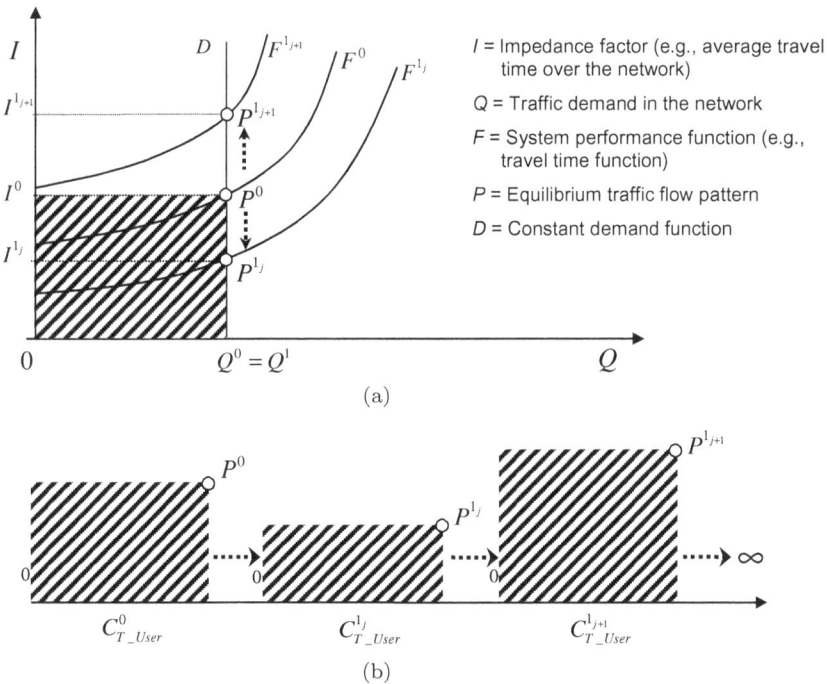

Figure 10.2 Concept for evaluating user cost saving: (a) network performance variation with different highway alternatives and (b) total user-cost variation with different highway alternatives

AASHTO (2010) for major highway construction projects. Additionally, since traffic patterns vary with time of day, different time periods during a day (AM, PM, and OFF peaks) are considered for more precise cost estimation. According to the AASHTO (2003), a model day of 18 hours is suggested, with traffic in six hours from 12 midnight to 6 AM added to the off-peak period. There are about 261 weekdays in a year. If we let H_{AM} and H_{PM} be AM and PM peak hours in a day, respectively, then the total number of AM and PM peak hours in a year becomes $309 H_{AM}$ and $309 H_{PM}$, respectively. The total number of off-peak hours/year becomes $(365 \times 18) - 261 \times (H_{AM} + H_{PM})$.

(A) *Travel Time Cost Saving*

The travel time cost is calculated based on the amount of time spent for traveling and the drivers' perceived value of time. Before elaborating on the travel time cost model, it is important to note that there are multiple user classes operating on the highway network, and each user class may have different trip purposes. In this monograph, it is assumed that two types of user classes (auto and truck) operate on the highway network, and they have different values of travel time with respect to different trip purposes (see Table 10.2). A more detailed trip purpose factor for each user class (such as, home-based work, home-based other, and non-home-based trips) is not considered here, since it may not be available in the initial stage of the highway project due to either time or money constraint (or both).

Table 10.2 Wage compensation rate for different trip purposes

Mode	Trip purpose	Percentage of wage compensation
Auto	Drive alone commute	50% of the wage rate
	Carpool driver commute	60% of the wage rate
	Carpool passenger commute	40% of the wage rate
	Personal	50% ∼ 70% of the wage rate
	Average	50% of the wage rate
Truck	In-vehicle and excess (waiting time) business	100% of total compensation

Source: AASHTO and US Department of Transportation (1997). The Value of Travel Time: Departmental Guidance for Conducting Economic Evaluations, Washington, D.C.

Table 10.3 Average wages, by industry (2000 US dollars)

Industry Type	Average Wage ($/hr)
All employees (Auto users)	$18.56
Truck drivers	$16.84

Source: National Income and Product Accounts (NIPA) of the United States, for 2000 (Department of Commerce, Bureau of Economic Analysis).

Table 10.4 Average vehicle occupancy for autos and trucks

Vehicle types	Average vehicle occupancy
Auto	1.550 (persons/vehicle)
Truck	1.144 (persons/vehicle)

Source: Oak Ridge National Laboratories, 1995 National Personal Travel Survey Databook, ORNL/TM-2001/248, Table 7.6, Oct. 2001.

Table 10.2 shows wage compensation rates suggested in AASHTO (2003) with respect to different trip purposes. With these guidelines, the unit travel time value for each user class (i.e., $\mathbf{v} = [v_{Auto}, v_{Truck}]$) can be simply calculated by multiplying the wage compensation rate (here we use average values shaded Table 10.2) by the corresponding average wage shown in Table 10.3.

Vehicle occupancy information is also required to evaluate travel time cost of highway users more precisely. Let $\mathbf{o}$ be the vector of the average vehicle occupancy for the auto and truck drivers in the traffic flows; $\mathbf{o} = [o_{Auto}, o_{Truck}]$. Table 10.4 presents average vehicle occupancy information from the National Personal Travel Survey (NPTS, 1995). Additionally, we define a traffic composition vector (denoted as $\mathbf{T}$) for autos and trucks in the traffic flows; $\mathbf{T} = [1 - T_{Truck}, T_{Truck}]$. The truck percentage ($T_{Truck}$) is needed as a model input, and thus the auto percentage in the traffic becomes $1 - T_{Truck}$.

We now estimate economic value of the travel-time cost savings based on (i) the values obtained from Tables 10.2 to 10.4 and (ii) traffic performance measures obtained from the traffic assignment process (for before and after a new highway development). The following steps show how the travel-time saving is estimated in the optimization process:

STEP 1: Update traffic volume and travel time on all highways in the network from the traffic assignment process

- AM peak: $x_{a^0}^{AM}, t_{a^0}^{AM}, x_{a^1}^{AM}, t_{a^1}^{AM}$
- PM peak: $x_{a^0}^{PM}, t_{a^0}^{PM}, x_{a^1}^{PM}, t_{a^1}^{PM}$ for all $a^0 \in A^0$ and $a^1 \in A^1$
- OFF peak: $x_{a^0}^{OFF}, t_{a^0}^{OFF}, x_{a^1}^{OFF}, t_{a^1}^{OFF}$

where A^0, A^1 = a set of arcs in the road network before and after a new highway is added, respectively; $a^0 \in A^0$, $a^1 \in A^1$ $\mathbf{v}$ = ravel time values for auto and truck, $\mathbf{v} = [v_{Auto}, v_{Truck}]$; v_{Auto} and v_{Truck} are unit travel time value (\$/hr) of auto and truck drivers, respectively; $x_{a^0}^{AM}, x_{a^0}^{PM}, x_{a^0}^{OFF}$ are average traffic flows in vehicles per hour on arc a^0 during AM, PM, and OFF peak hours, respectively, before a new highway is added in the network. $t_{a^0}^{AM}, t_{a^0}^{PM}, t_{a^0}^{OFF}$ are average travel time (hours) that vehicles spent on arc a^0 during AM, PM, and OFF peak hours, respectively, before a new highway is added in the network. $x_{a^1}^{AM}, x_{a^1}^{PM}, x_{a^1}^{OFF}$ are average traffic flows in vehicles per hour on arc a^1 during AM, PM, and OFF peak hours, respectively after a new highway is added in the network. $t_{a^1}^{AM}, t_{a^1}^{PM}, t_{a^1}^{OFF}$ are average travel time (hours) that vehicles spent on arc a^1 during AM, PM, and OFF peak hours, respectively after a new highway is added in the network. Note that $x_{a^0}^{AM}, x_{a^0}^{PM}, x_{a^0}^{OFF}, t_{a^0}^{AM}, t_{a^0}^{PM}, t_{a^0}^{OFF}$ are computed ONLY ONCE at the beginning of the optimization process with the HAO model.

STEP 2: Calculate total travel time cost over the network

$$C_{T\_B}^0 = \sum_{a^0 \in \mathbf{A}^0} \left(\begin{bmatrix} x_{a^0}^{AM} \\ x_{a^0}^{PM} \\ x_{a^0}^{OFF} \end{bmatrix} \cdot \begin{bmatrix} t_{a^0}^{AM} \\ t_{a^0}^{PM} \\ t_{a^0}^{OFF} \end{bmatrix} \cdot \begin{bmatrix} H_{TAM} \\ H_{TPM} \\ H_{TOFF} \end{bmatrix} \right) [\mathbf{v} \cdot \mathbf{T} \cdot \mathbf{o}] \quad (10.9)$$

$$C_{T\_B}^1 = \sum_{a^1 \in \mathbf{A}^1} \left(\begin{bmatrix} x_{a^1}^{AM} \\ x_{a^1}^{PM} \\ x_{a^1}^{OFF} \end{bmatrix} \cdot \begin{bmatrix} t_{a^1}^{AM} \\ t_{a^1}^{PM} \\ t_{a^1}^{OFF} \end{bmatrix} \cdot \begin{bmatrix} H_{TAM} \\ H_{TPM} \\ H_{TOFF} \end{bmatrix} \right) [\mathbf{v} \cdot \mathbf{T} \cdot \mathbf{o}] \quad (10.10)$$

where $C_{T\_B}^0, C_{T\_B}^1$ = total travel time cost in the base year (\$/yr) over the road network before and after a new highway is added, respectively;

$H_{TAM}, H_{TPM}, H_{TOFF}$ = total number of AM, PM, and OFF peak hours in a year, respectively; $H_{TAM} = 261H_{AM}, H_{TPM} = 261H_{PM}, H_{TOFF} = 365 \times 18 - 261(H_{PM} + H_{PM}); H_{AM}, H_{PM}, H_{OFF}$ = AM, PM, and OFF peak hours in a day, respectively; Other terms are defined previously.

STEP 3: Calculate present value of total travel time cost saving

$$\Delta C_T = (C^0_{T\_B} - C^1_{T\_B}) \left(\frac{e^{(r_t-\rho)n_y} - 1}{r_t - \rho} \right) \tag{10.11}$$

where ΔC_T = present value of total travel time cost saving after a new highway development in the existing road network for the analysis period n_y; n_y = total analysis period (the number of years); ρ = assumed interest rate (decimal fraction); r_t = annual growth rate of traffic over the road network (decimal fraction)

Note that intersections may entail additional travel time (delay) besides the arc travel time (t_a), while providing right-of ways to all turning movements entering the intersection. Thus, if there is any intersection in the highway network, the intersection delay (d_a) may also be considered in the travel time cost estimation procedure presented earlier. Intersection-delay functions used in the bi-level HAO model are presented in Section 10.5.2.

(B) *Vehicle Operating Cost Saving*

Another user cost component considered in the model is the 'vehicle operating cost' that can be directly perceived by drivers (the network users) as an out-of-pocket expense incurred while operating vehicles. This may include fuel and oil, maintenance, tire wear, and vehicle depreciation costs. However, since the vehicle depreciation cost is not sensitive to network configuration with different highway alternatives to be added, only fuel consumption and vehicle maintenance (including tire wear cost) costs, which are the most dominating and sensitive ones, are considered in this analysis. Generally, the vehicle operating cost can be calculated on a per vehicle-mile basis; link distance as well as equilibrium link traffic information (such as flow, travel time, and speed on each link), which are obtained from the traffic assignment process, are used to estimate the vehicle operating cost. The fuel consumption (efficiency) indices, expressed in gallons per mile at different average operation speeds, for the two different user-classes (auto and truck) are presented in Table 10.5, and their fuel prices

Table 10.5 Fuel consumption rates for autos and trucks

Speed	Gallons/mile	
	Auto	Truck
5 mph	0.117	0.503
10 mph	0.075	0.316
15 mph	0.061	0.254
20 mph	0.054	0.222
25 mph	0.050	0.204
30 mph	0.047	0.191
35 mph	0.045	0.182
40 mph	0.044	0.176
45 mph	0.042	0.170
50 mph	0.041	0.166
55 mph	0.041	0.163
60 mph	0.040	0.160
65 mph	0.039	0.158

Source: Meyer (2016). "Chapter 4: Environmental and Energy Considerations," in Transportation Planning Handbook, 4th Edition. Institute of Transportation Engineers.

Table 10.6 Auto and truck fuel prices and maintenance and tire costs

Category	Auto	Truck
Fuel	$p_{F_{Auto}} = \$2.73/\text{gallon}$	$p_{F_{Truck}} = \$2.67/\text{gallon}$
Maintenance and Tires*	$m_{Auto} = \$0.04//\text{mile}$	$m_{Truck} = \$0.05/\text{mile}$

*Source: American Automobile Association, *Your Driving Costs*, 1999 Edition. Data for a popular model of each type listed with ownership costs based on 60,000 miles before replacement. Adjusted to 2002 dollars by ECONorthwest.

(dollars per gallon) and average maintenance and tire costs are shown in Table 10.6. These are also required inputs for estimating the vehicle operation cost.

Then, the unit vehicle-operating cost can be estimated with the sum of (i) unit fuel consumption costs, calculated by multiplying the fuel consumption and the fuel price, and (ii) maintenance and tire costs. As shown in Table 10.6, we use $2.73 and $2.67 per gallon for fuel prices of auto and truck modes, respectively, and $0.040 and $0.050 per mile are used for representing their unit vehicle maintenance and tire wear costs. Since the fuel efficiency varies depending on different modes (here autos and trucks) as well as their operating speeds, we use regression functions obtained from the data provided in Table 10.5. Equations (10.12) and (10.13) show unit fuel-consumption functions for the two different modes resulting from the regression analysis.

$$f_{a\_Auto} = 0.194(S_{a\_Auto})^{-0.401} \qquad (10.12)$$

$$f_{a\_Truck} = 0.194(S_{a\_Truck})^{-0.429} \qquad (10.13)$$

where S_{a_Auto}, S_{a_Truck} = average speeds (miles/hr) of autos and trucks on arc a, respectively.

The total vehicle operation cost for users' travels on the network can be calculated by multiplying the unit value of the vehicle operating cost (given per vehicle-mile for corresponding link) by the arc travel distance and corresponding link traffic volume. Note that the total vehicle operating cost in the given highway network may increase after the new highway construction, since the construction scenario entails more links (road segments), and thus possibly more overall travel in the study area. However, "the average vehicle operating cost per vehicle-mile over the study area would be expected to decrease with the improved network; in other words, under improved travel conditions, the per-unit cost of travel decreases" (Clifton and Mahmassani, 2004). The following steps show how the vehicle operating cost saving is estimated in our optimization model:

STEP 1: Update alignment length, traffic volume, and unit vehicle operating cost of all highways from the traffic assignment process

- AM peak: $x_{a^0}^{AM}, \mathbf{f}_{a^0}^{AM}, x_{a^1}^{AM}, \mathbf{f}_{a^1}^{AM}$
- PM peak: $x_{a^0}^{PM}, \mathbf{f}_{a^0}^{PM}, x_{a^1}^{PM}, \mathbf{f}_{a^1}^{PM}$ for all $a^0 \in \mathbf{A}^0$ and $a^1 \in \mathbf{A}^1$
- OFF peak: $x_{a^0}^{OFF}, \mathbf{f}_{a^0}^{OFF}, x_{a^1}^{OFF}, \mathbf{f}_{a^1}^{OFF}$

where L_{a^0}, L_{a^1} = length of arc a^0 and a^1, respectively; $a^0 \in \mathbf{A}^0$ and $a^1 \in \mathbf{A}^1$; $\mathbf{A}^0$, $\mathbf{A}^1$ = a set of arcs in the road network before and after a new highway is added; $\mathbf{f}_{a^0}^{AM}, \mathbf{f}_{a^0}^{PM}, \mathbf{f}_{a^0}^{OFF}$ = vector representations of unit vehicle operating cost for traffic (autos and trucks) on arc a^0 during AM, PM, and OFF peak hours, respectively, before a new highway is added in the road network; $\mathbf{f}_{a^1}^{AM}, \mathbf{f}_{a^1}^{PM}, \mathbf{f}_{a^1}^{OFF}$ = vector representations of unit vehicle operating cost for traffic (autos and trucks) on arc a^1 during AM, PM, and OFF peak hours, respectively, after a new highway is added in the road network; Note that $x_{a^0}^{AM}$, $x_{a^0}^{PM}$, $x_{a^0}^{OFF}$, $f_{a^0}^{AM}$, $f_{a^0}^{PM}$, $f_{a^0}^{OFF}$ are computed ONLY ONCE at the beginning of the optimization process with the HAO model.

STEP 2: Calculate total vehicle operating cost over the network

$$C_{V\_B}^0 = \frac{1}{5280} \sum_{a^0 \in A^0} \left(\begin{bmatrix} x_{a^0}^{AM} \\ x_{a^0}^{PM} \\ x_{a^0}^{OFF} \end{bmatrix} \cdot \begin{bmatrix} f_{a^0}^{AM} \cdot \mathbf{T} \\ f_{a^0}^{PM} \cdot \mathbf{T} \\ f_{a^0}^{OFF} \cdot \mathbf{T} \end{bmatrix} \cdot \begin{bmatrix} H_{TAM} \\ H_{TPM} \\ H_{TOFF} \end{bmatrix} \right) \quad (10.14)$$

$$C_{V\_B}^1 = \frac{1}{5280} \sum_{a^1 \in A^1} \left(\begin{bmatrix} x_{a^1}^{AM} \\ x_{a^1}^{PM} \\ x_{a^1}^{OFF} \end{bmatrix} \cdot \begin{bmatrix} f_{a^1}^{AM} \cdot \mathbf{T} \\ f_{a^1}^{PM} \cdot \mathbf{T} \\ f_{a^1}^{OFF} \cdot \mathbf{T} \end{bmatrix} \cdot \begin{bmatrix} H_{TAM} \\ H_{TPM} \\ H_{TOFF} \end{bmatrix} \right) \quad (10.15)$$

where $C_{V\_B}^0$, $C_{V\_B}^1$ = total vehicle operating cost in the base year (\$/yr) over the road network before and after a new highway is added, respectively.

Other terms are defined previously.

STEP 3: Calculate present value of total vehicle operating cost saving

$$\Delta C_V = (C_{V\_B}^0 - C_{V\_B}^1) \left(\frac{e^{(r_t - \rho)n_y} - 1}{r_t - \rho} \right) \quad (10.16)$$

where ΔC_V = present value of total vehicle operating cost saving after a new highway development in the existing road network for the analysis period n_y

Other terms are defined previously.

An at-grade intersection may also increase vehicle fuel cost to the highway users since it delays and stops vehicles. Thus, if any at-grade intersections are included in the given highway network, the vehicle fuel cost associated with the intersections should also be included in the total vehicle operating cost estimated for the network.

10.2.3 Penalty and Environmental Costs

As discussed in Chapter 5, the penalty function (C_P) can also be included in the objective function of the bi-level HAO model to smoothly guide the search for the highway alignment optimization process, while ensuring a high penalty is given to a solution alignment which affects the area of a land parcel more than its pre-defined maximum allowable limit. The penalty function can also be used where a generated solution alignment does not satisfy specified design constraints; for instance, the penalty may be assigned if the new highway alignment does not accommodate the required minimum curve length.

10.3 Lower Level of Bi-level HAO

It is assumed that the network drivers adjust their travel paths with response to various network configurations due to the addition of different candidate alignments. The traffic assignment problem, which is considered as the lower-level problem in the model, is designed to represent such a traffic effect in the evaluation process. Equilibrium traffic flows for the changing highway network are estimated from the traffic assignment process in the model, and they are ultimately used for computing costs associated with user travels in the upper-level problem.

10.3.1 User and System Optimal Traffic Assignment Problems

Typically, two assignment principles are used in the assignment problem for representing interaction between supply (network design) and demand (network users) actions. These are user optimal (UO) and system optimal (SO) principles:

- *User optimal (UO): origin/destination flows are assigned to possible paths in the network with minimum travel time.*
- *System Optimal (SO): origin/destination flows are assigned such that total travel time in the network is minimized.*

The static UO assignment problem is to find the arc flows, $\mathbf{x}$, that satisfy the user equilibrium criterion when all the origin-destination entries, $q_{r,s}$ for all r, s ($r \neq s$) have been appropriately assigned (Sheffi, 1984). This equilibrium arc-flow pattern can be obtained by solving Eq. (10.2), originally proposed by Beckmann *et al.* (1956), subject to three types of constraints: (i) flow conservation constraints, (ii) non-negativity constraints, and (iii) incident relationships between arc and path flows. Note that the UO objective function (i.e., Eq. (10.2) is strictly convex everywhere (in $\mathbf{x}$)

for the static traffic assignment problem (Sheffi, 1984); thus, it has a unique solution.

The SO assignment problem is to find the arc flows, $\mathbf{x}$ that minimize total travel time of the network subject to the same constraints as in the UO assignment problem. The SO objective function, Eq. (10.3) is also strictly convex in $\mathbf{x}$ for the same criteria as in the UO problem and may be rewritten as the following formulation (Sheffi, 1984 and Thomas, 1991):

$$\text{Minimize } Z_{LL} = \sum_a \int_0^{x_a} \left(\frac{d(x_a t_a(x_a))}{dx_a} \right) dx_a \quad \text{for system optimum}$$

(10.17)

The solution from the SO assignment problem with Eq. (10.17) indicates that all used paths have equal and minimum marginal-travel times between any O/D pair, while that from the UO problem indicates that all used paths have equal and minimum travel time between any O/D pair (Thomas, 1991). Note that a decision on which assignment principle (between UO and SO) is used is user-specifiable in the proposed optimization model.

A static (and deterministic) traffic assignment method is adopted in the bi-level HAO model since it is commonly used in many planning applications. The assignment method, although it has a limitation in capturing traffic phenomena such as the propagation of shockwaves and queue spillovers, is widely used by agencies for planning applications from infrastructure structure improvement to traffic maintenance and congestion management.

The convex combinations method, originally developed by Frank and Wolfe in 1956, has been widely used for solving quadratic programming problems with linear constraints. The method is also known as the Frank-Wolfe algorithm and is very useful for general application of the static traffic assignment problems; both the UO and SO traffic assignment principles are applicable to the well known iterative algorithm. Starting with a feasible solution (i.e., a set of arc flows, x_{a^0} for $a^0 \in \mathbf{A}^0$), the Frank-Wolfe algorithm converges after a finite number of iterations. The basic Frank-Wolfe algorithm may be found in Sheffi (1984) and Thomas (1991).

10.3.2 Determination of Traffic Reassignment

Recall that the bi-level optimization approach is designed for (i) updating the configuration of a road network after a new highway alignment is generated, (ii) finding equilibrium traffic flows in the updated network from the traffic assignment process, and (iii) evaluating the total cost of the new highway including the user and agency cost items. It should be noted that the bi-level optimization approach may not be efficient in cases when the traffic assignment results for the networks updated with different highway alternatives are quite similar. For instance, the difference in traffic volumes which would operate on the highway alternatives may be negligible although their start and end points as well as horizontal (and even vertical) alignments generated from the model significantly differ. In such a case, the traffic reassignment (i.e., finding equilibrium traffic flows for every network updated with new highway alternatives) is wasteful. Thus, the following preprocessed traffic assignment procedure is added in the model for determining whether the bi-level optimization feature is needed during the alignment search process. The preprocessed traffic assignment is intended to accelerate the alignment evaluation procedure and enhance the model's computational efficiency accordingly. Note that this procedure (as described in the following) is preprocessed with sample solutions generated at early stages of the highway alignment optimization, including the initial population stage.

Preprocessed Traffic Assignments

Step 1: Generate sample highway alternatives including straight and curved alignments.

Step 2: Update configuration of the road network with the new highway generated.

Step 3: Find equilibrium traffic volumes distributed to the updated road network from the traffic assignment process.

Step 4: Calculate the coefficient of variation (CoV) of the estimated traffic volumes to check whether the traffic assignment results are relatively consistent for the sample highway alternatives (note that a small CoV indicates that the assignment results are relatively consistent).

- If the CoV is within the range of a threshold value:
  - → Stop the traffic reassignment for all alternatives generated over the successive generations. Instead, the results of the traffic assignments

with the sample alternatives will be used for estimating the user costs of the other alternatives.

- Otherwise:

  → The traffic reassignment will be processed for all alternatives generated over the successive generations for estimating distributed traffic and the user cost.

10.4 Bi-level HAO Model Structure

In the bi-level HAO model, GAs with a number of specialized genetic operators is used for searching optimized 3D highway alignments. Highway alignments are represented as chromosomes, and each chromosome has a set of genes defined with xyz coordinates of PI's (refer to Figure 4.1 in Chapter 4). In addition, a Geographic Information System (GIS) and an equilibrium traffic assignment (TA) are integrated with the GA to realistically evaluate the highway alignments generated. During the alignment search process, the GA, GIS, and TA communicate with each other by transmitting model inputs and outputs (see Figure 10.3)). To create a horizontal alignment, PI's of a new highway alignment are first generated from the GA. The successive PI's are then connected with straight lines ("tangents"), and circular curves are fitted to connect the tangents. Note that transition curves can also be added in between the tangents and circular curves if high-level design standards are required. The corresponding vertical alignment is also determined by fitting parabolic curves to graded-tangents at PI's, while the horizontal alignment is being created.

After the 3D highway alignment is generated and the road network is updated with it, the equilibrium TA is processed. Results of the TA process are returned to the GA for evaluating the user cost components. Spatial information of the generated highway alignment is also transmitted from the GA to the GIS, and thus alignment's right-of-way cost, environmental impact, and socio-economic impact are evaluated in the GIS, while the other alignment-sensitive costs (e.g., earthwork and pavement costs) are computed in the GA.

In the bi-level HAO model, three main types of inputs are needed for optimizing the 3D highway alignments. First, the design specification, normally defined based on AASHTO design standards (AASHTO, 2011), is

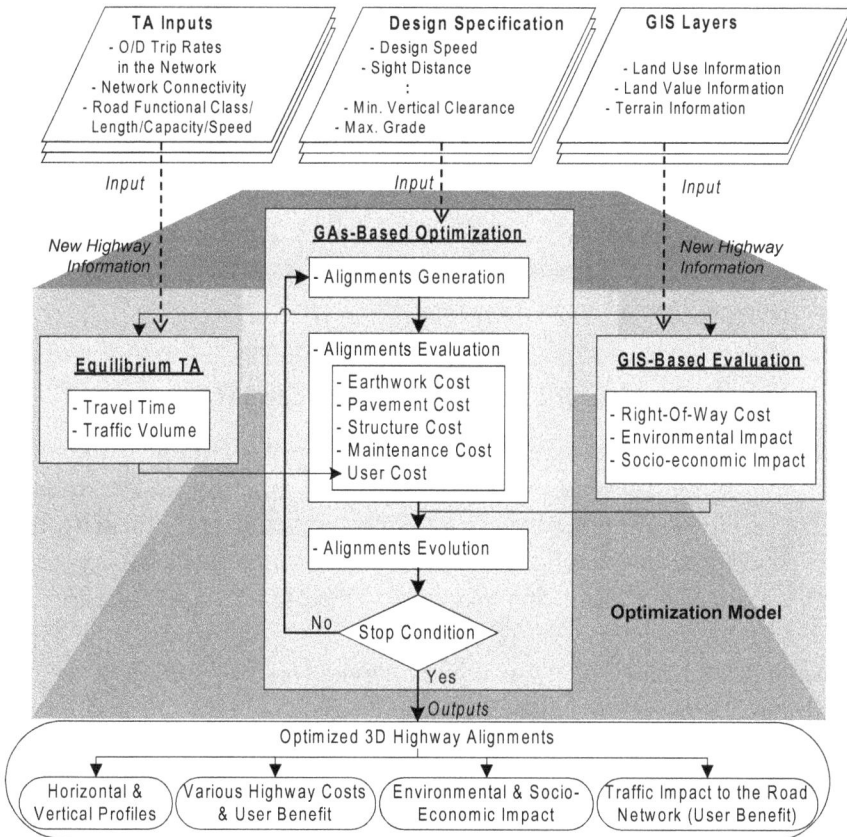

Figure 10.3 Basic model structure of bi-level HAO model

needed for generating the highway alignments, and the alignments are evaluated based on several unit costs (e.g., unit pavement cost and unit earthwork
cost) defined by the model users. Second, the GIS inputs are essential for
computing right-of-way cost of the new highway and evaluating its environmental and socio-economic impacts. The model users can also express their
preferences, by specifying their areas of interest and untouchable areas in
the GIS layers. Lastly, information about current and future traffic as well
as about the existing road network is also necessary for the TA process.
O/D trips, functional class, length, and capacity of the roads in the existing network as well as the network connectivity are included in this input
class. By comparing the TA results for the networks before and after the

new highway addition, the users cost savings (e.g., travel time and vehicle operation cost savings) can be estimated.

Various practical and quantitative results of the optimized alignments are provided as model outputs. The model output includes horizontal and vertical profiles of the optimized highway alignments, traffic and environmental impacts on the study area, and breakdown of the total highway cost. These are quite useful to the decision-makers for identifying and refining new highways. Note that the model outputs include a graphical view of the optimized alignments on a GIS map.

10.5   Inputs Required for Lower Level of Bi-level HAO

Large amounts of input data are required to perform a traffic assignment analysis for a highway network. These include typically (i) the physical layout of the highway network (e.g., highway type, length, capacity, and free-flow speed and location of highway junction points), (ii) trip rates between origin and destination nodes (i.e., O/D trip matrix) of the network, and (iii) travel time functions (known as link performance or volume-delay functions) for estimating travel time of the network travelers. This section describes those input requirements.

10.5.1   Highway Network and O/D Trip Matrix

In the traffic assignment problem, a real highway network can be represented as a directed graph consisting of a finite set of nodes (or points or vertices) and pairs of which joined by one or more arcs (or links) in the network. Figures 10.4 and 10.5 show representation of a typical highway network (before and after a new highway addition) used for the traffic assignment process. In the figures, numbers marked with an italic font stand for a set of arcs, and those with boldface depict a set of nodes.

Normally, two subsets of nodes are used for the network representation. The first one is a set of origin and destination points (known as centroids) at which all trips are assumed to start and finish. The other one is a set of junction nodes representing points at which highways intersect each other (e.g., interchanges and intersections) or points at which physical nature of a highway is remarkably changed (e.g., highway-capacity-increase points due to increase in number of lanes). Dummy nodes may also be used to more

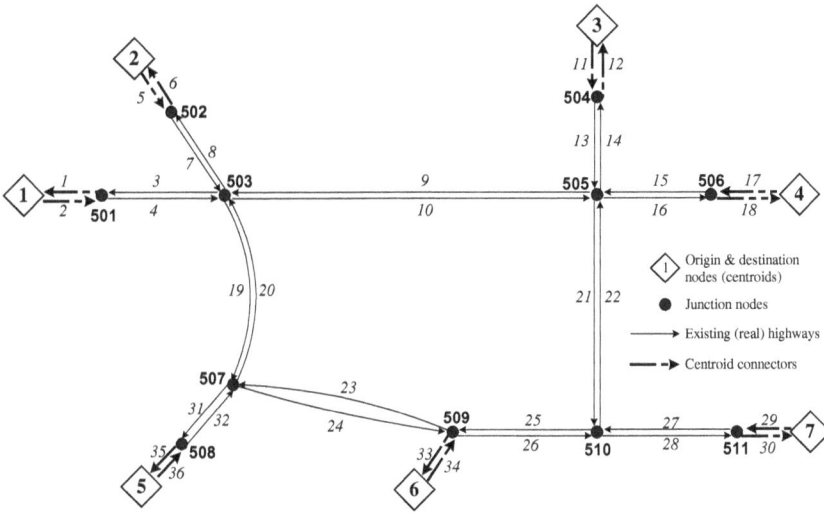

Figure 10.4 An example road network before a new highway addition

realistically represent a highway junction node by breaking it up into several dummy points; typically more dummy nodes are used for more microscopic network representation.

Arcs are basically one-way sections of highways, and are typically iden-tified by their start and end points (called initial nodes and final nodes, respectively). Two types of arcs are normally used for representing the highway network: (i) centroid connectors and (ii) highway arcs. Centroid connectors are not actual roads but conceptual representations of arcs that connect the centroids (trip origin and destination points) and the highway junction nods. Highway arcs connect the highway junction nodes, and can be classified into several sets of categories based on their speeds and capac-ities and access control types designed.

In the traffic assignment of the lower level problem, four types of high-way arcs are used with different levels of the design characteristics; these are freeways, expressways, arterials, and collectors (refer to Table 10.7). Additionally, dummy arcs may also be used for representing the highway network by connecting dummy nodes at both of their ends (readers may refer to Figure 6.4). More detailed discussion of the node and arc (link) rep-resentation for traffic assignment problems may be found in many related studies such as Sheffi (1984) and Thomas (1991).

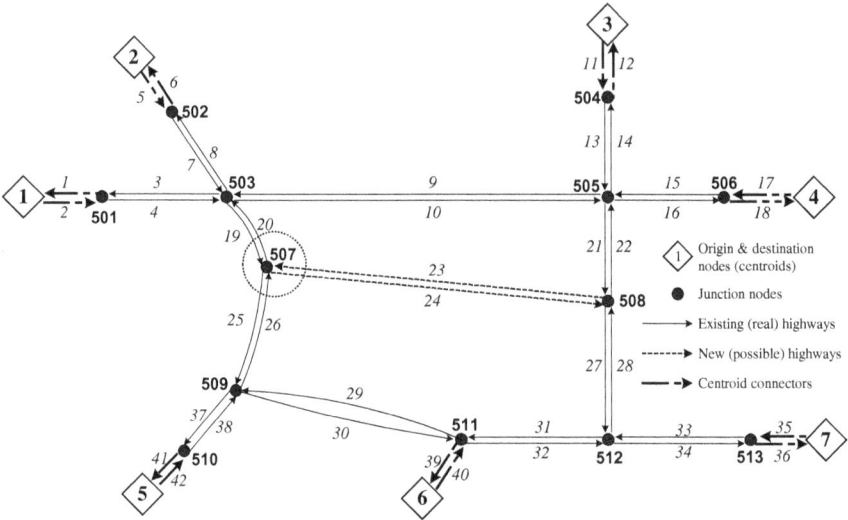

Figure 10.5   An example road network after a new highway addition

Let us now suppose that a new highway alignment is generated from the *Highway Alignment Generation Procedure* (Section 4.2.1), connecting in the middle of existing highways 19–20 and 21–22 in the highway network shown in Figure 10.4. Then, the network might be modified as the one shown in Figure 10.5. Note that the newly connected alignment has its realistic horizontal and vertical profiles with various design specifications although it is just represented with two straight lines (arcs) in the figure. In Figure 10.5, the numbers of highway arcs and junction-nodes are increased (as compared to the network shown in Figure 10.4) due to the highway addition process. Arc (ID) numbers and properties (e.g., lengths of highway arcs *19, 20, 21, 22, 23, 24, 25, 26, 27, and 28* and locations of highway nodes **507** and **508**) are also newly updated from that process.

Table 10.7 presents an example input layout of the arcs and arc properties (e.g., road type, length, capacity, speed, exit type, etc.) used in the traffic assignment process for the network updated from the highway addition. In the table, shaded rows indicate (i) the arcs of the new alignment added to the existing network and (ii) the arcs whose properties are updated from the highway addition. Note that they are iteratively updated whenever the new alignments are generated during the optimization process.

Table 10.7 Example input layout of a highway network for the assignment process

Arc number	Initial node	Final node	Road type	Length (feet)	Number of lanes	Capacity (vphpl)	Free-flow speed (mph)	Exit type	Cycle length (sec)	Effective green (sec)
1	501	1	Centroid	0	2	99, 999	65	0	—	—
2	1	501	Centroid	0	2	99, 999	65	0	—	—
3	503	501	Freeway	13, 411	2	2, 200	65	0	—	—
4	501	503	Freeway	13, 464	2	2, 200	65	0	—	—
5	2	502	Centroid	0	2	99, 999	50	0	—	—
6	502	2	Centroid	0	2	99, 999	50	0	—	—
7	502	503	Arterial	16, 632	2	1, 800	50	0	—	—
8	503	502	Arterial	16, 648	2	1, 800	50	0	—	—
·	·	·	·	·	·	·	·	·	·	·
·	·	·	·	·	·	·	·	·	·	·
·	·	·	·	·	·	·	·	·	·	·
19	503	507	Arterial	8, 131	2	1, 800	45	0	—	—
20	507	503	Arterial	8, 158	2	1, 800	45	0	—	—
21	505	508	Freeway	11, 616	2	2, 200	65	0	—	—

(*Continued*)

Table 10.7 *(Continued)*

Arc number	Initial node	Final node	Road type	Length (feet)	Number of lanes	Capacity (vphpl)	Free-flow speed (mph)	Exit type	Cycle length (sec)	Effective green (sec)
22	508	505	Freeway	11,616	2	2,200	65	0	—	—
23	508	507	Arterial	18,110	2	1,800	65	0	—	—
24	507	508	Arterial	18,110	2	1,800	65	0	—	—
25	507	509	Arterial	11,405	2	1,800	45	0	—	—
26	509	507	Arterial	11,405	2	1,800	45	0	—	—
27	508	512	Freeway	9,821	2	2,200	45	0	—	—
28	512	508	Freeway	9,800	2	2,200	45	0	—	—
.	.	.	.	.	.	.	.	.	.	.
37	509	510	Arterial	7,530	2	1,800	45	0	—	—
38	510	509	Arterial	7,530	2	1,800	45	0	—	—
39	511	6	Centroid	0	2	99,999	50	0	—	—
40	6	511	Centroid	0	2	99,999	50	0	—	—
41	510	5	Centroid	0	2	99,999	45	0	—	—
42	5	510	Centroid	0	2	99,999	45	0	—	—

Exit type: 0 = grade-separated interchange, 1 = signalized intersection
2 = stop-controlled intersection, 3 = roundabout

The exit type in Table 10.7 is used for indicating the type of access-control designed at the final node of each highway arc (e.g., grade separated interchange and signalized intersection). For instance, if a highway arc is approaching a signalized intersection, additional information required for reflecting the intersection effect (such as cycle length and effective green time for traffic using that arc) would be needed for calculating the arc capacity. Note that in the traffic assignment of the lower level problem, interchange junction-effects are assumed to be negligible; thus, an interchange can be represented with a single junction node (e.g., node **507** in Figure 10.5), while the at-grade intersection may be represented with several dummy nodes and arcs as shown in Figure 10.6.

Figure 10.6 shows a microscopic representation of junction node **507**, which is one of the alignment endpoints of the new highway added to the network. Such a detailed sub-network may be used in the network representation if a junction node is considered as an intersection node or if traffic delay at that node is particularly important in the traffic assignment process. Intersection volume-delay functions may be employed for estimating travel time of travelers on the dummy arcs while travel time functions (e.g., BPR functions) are used for travel time on other regular highway arcs.

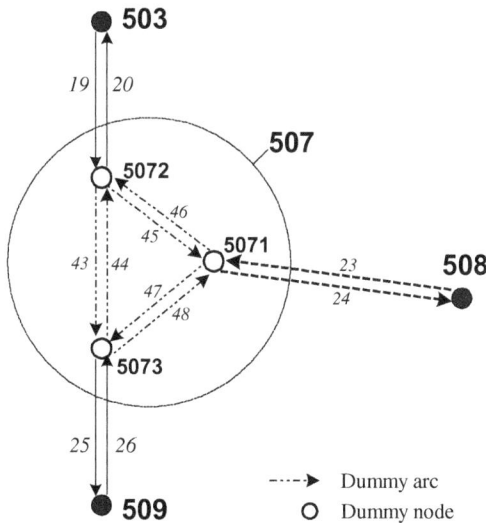

Figure 10.6 Microscopic representation of a highway node

Table 10.8 Example input layout of O/D matrix

Origin node	Destinatioin node	Average traffic volume (veh/hr)	Peak hour factor		Peak duration (hr)	
			AM peak	PM peak	AM peak	PM peak
1	*2*	85	0.9	0.8		
1	*3*	70	1.1	1		
1	*4*	600	1.3	0.7		
1	*5*	200	1.3	0.7		
1	*6*	300	1.4	0.6		
1	*7*	1000	1.3	0.5		
.	.	.	.	.		
.	.	.	.	.		
.	.	.	.	.		
4	*1*	600	0.6	1.4		
4	*2*	400	0.7	1.3		
4	*3*	350	0.8	0.8	3	3
4	*5*	80	1.2	0.7		
4	*6*	600	1.4	0.7		
4	*7*	600	1.3	0.8		
.	.	.	.	.		
.	.	.	.	.		
.	.	.	.	.		
7	*1*	1000	0.5	1.3		
7	*2*	400	0.6	1.4		
7	*3*	300	0.9	0.9		
7	*4*	800	0.7	1.3		
7	*5*	200	1	1		
7	*6*	200	1	1		

Besides the network information stated earlier, specification of origin and destination (O/D) trip matrix is also required to perform the traffic assignment process. An example layout of input O/D trip matrix used in the assignment process for the example network shown in Figure 10.5 is illustrated in Table 10.8. As shown in the table, basically trip rates of origin-destination pairs and their peak hour durations are required to construct the input O/D matrix.

10.5.2 Travel Time Functions

Many travel-time functions (which are also known as link-performance functions or volume-delay models) have been developed for estimating link (highway) travel time in the transportation system planning stage. These include the equation developed by the United States Bureau of Public Roads (BPR, 1964) and its modified versions (Singh, 1995; Dowling *et al.*, 1998; AASHTO, 2003), Webster's (1958) volume-delay model, Highway Capacity Manual (HCM) by Transportation Research Board (TRB, 2010) methods, Davidson's (1966, 1978) model, TRANSYT-7F models (Wallace *et al.*, 1998), and Akcelik's (1991) model. Among them, simplified functions that are often (practically) applied to the traffic assignment problem for estimating travel time on highway segments are the 2003 BPR (AASHTO, 2003) model and Akcelik's (1991) model. For estimating delay time on at-grade intersections and roundabouts, the HCM (2010) model is useful.

Link Performance Functions for Highway Travel Time Estimation

The BPR function has been extensively updated until recently, and currently AASHTO (2003) proposes different model parameters in terms of various road-types (e.g., freeways and arterials; refer to Table 10.9). This travel time function can characterize traffic volume-delay relationships with a simple algebraic form that is easy to remember and work with other mathematical models. The 2003 BPR model can replicate most observed patterns of delay by selecting the proper combination of free-flow travel time and the two parameters, α and β. The basic formulation of the BPR model is:

$$t_a(x_a) = t_a^f \left[1 + \alpha \left(\frac{x_a}{c_a'} \right) \right] \qquad (10.18)$$

Table 10.9 Typical BPR function parameters

BPR parameters	Freeway	Expressway	Arterial	Collector
Urban				
Free-flow speed (mph)	55	45	30	25
α	0.1	0.1	0.05	0.075
β	10	10	10	10
Rural/Suburban				
Free-flow speed (mph)	65	55	45	40
α	0.1	0.1	0.05	0.075
β	10	10	10	10

Source: American Association of State Highway and Transportation Officials (AASHTO): User Benefit Analysis for Highways (2003).

where t_a = travel time (hr/mile) predicted for traffic on arc a; $a \in A$; x_a = traffic flow on arc a; $a \in A$; A = a set of arcs in the road network; t_a^f = free-flow travel time (hr/mile) on arc a; $t_a^f = L_a/S_a^f$; c_a' = practical capacity of arc a; L_a = Length of arc a; $a \in A$; S_a^f = free flow speed of arc a; $a \in A$; α, β are model parameters; refer to Table 10.9.

Note that the practical capacity of arc a (denoted as c_a' in the earlier equation) does not equal to its actual capacity (c_a) which represents maximum possible flows that can pass through the arc in a given time; c_a' can be defined as 80% of the actual arc capacity (Dowling *et al.*, 1998). Table 10.9 presents the BPR parameter values used in Eq. (10.18). In the equation, the parameter α determines the ratio of free-flow speed to the speed at capacity, and parameter β determines how abruptly the BPR curve drops from the free-flow speed.

Akcelik (1991) proposed a modified version of Davidson's (1966 and 1978) volume-delay model for properly using it in transport planning purposes. He avoided the limitations of the steady-state form of Davidson's model by developing a different functional form that uses the free-flow travel time and queuing delay terms in an explicit way. The following relation shows the time-dependent form of Akcelik's model:

$$t_a(x_a) = t_a^f + \left\{ 0.25\dot{T} \left[\left(\frac{x_a}{c_a} - 1 \right) + \sqrt{ \left(\frac{x_a}{c_a} - 1 \right)^2 + \left(\frac{8J_A}{c_a\dot{T}} \right) \left(\frac{x_a}{c_a} \right) } \right] \right\}$$

(10.19)

where $\dot{T}$ = duration; of traffic flow; typically 1 hour; c_a = capacity of arc a; $a \in A$; J_A = delay parameter; refer to Table 10.10.
Other terms are defined previously.

Equation (10.19) implies that the travel time estimate is the sum of the free-flow travel time along the highway (the first term) and delay due to queuing (the second term); the delay is equal to the average overflow queue divided by the capacity, c_a (refer to Akcelik, 1981). Table 10.10 shows example values of the delay parameter (J_A) suggested by Akcelik (1991). Note that although either the 2003 BPR model or Akcelik's (1991) model is applicable for predicting arc travel time in the traffic assignment process, we adopt the former travel time function because of its simpler mathematical form. A good comparison of those two models is provided in Dowling *et al.* (1998).

Table 10.10   Example delay parameters suggested by Akcelik (1991)

Facility type	Capacity (vphpl)	Free-flow speed		J_A	t_a^c / t_a^f
		(kph)	(mph)		
Freeway	2000	120	75	0.1	1.587
Expressway	1800	100	62	0.2	1.754
Arterial	1200	80	50	0.4	2.041
Collector	900	60	37	0.8	2.272
Local Street	600	40	25	1.6	2.439

t_a^c = travel time on arc a when $c_a = x_a$

Volume-Delay Models for Intersection Delay Estimation

The volume-delay models of signalized and stop-controlled intersections and roundabouts introduced in the Highway Capacity Manual (HCM) by Transportation Research Board (TRB, 2010) may be employed to estimate the intersection delays in the traffic assignment process. However, because their complex functional forms may cause a significant computational burden, the models may be used only when there are critical intersections which require detailed delay estimation a given highway network. Note that the volume-delay functions must be used numerous times in the traffic assignment analysis; furthermore, the assignment is processed whenever the candidate alignment is generated in the proposed network model. A simple intersection delay model, such as Webster's (1958) model may be considered for relaxing the computational complexity. However, since Webster's model does not cover oversaturated conditions,[1] in which intersection demand exceeds capacity, it is unsuitable for the assignment process.

The critical intersection may be represented with several dummy nodes and arcs, as shown in Figure 10.6, and the HCM models can be applied to the dummy arcs. For the other cases (e.g., interchanges or less important intersections connected between highway arcs), either the simple 2003 BPR model or Akcelik model can be used for the assignment process by a default. Intersection-delay models adopted in the bi-level HAO model can be found HCM 2010.

---

[1] In Webster's equation, the intersection-delay reaches infinity when the demand equals to capacity.

10.6 Summary and Future Work

Determination of a new highway location and alignment (including its geometric design, cost-benefit analysis, and analysis of its impacts to the existing land-use system) is a very complex and challenging problem due to the large number of conflicting factors that must be resolved, the great amount and variety of information that must be compiled and processed, and the numerous evaluations that must be performed. The process of evaluating even one candidate alternative with existing methods is so expensive and time consuming that typical studies can only afford to evaluate very few alternative alignments. In this chapter, we seek realistic highway alignments that best improve the existing roadway system, while considering their geometric designs, various costs associated with road construction, and environmental impacts to the study area. To solve such a complex optimization problem, a bi-level highway alignment optimization method is developed. In the bi-level model structure, the upper-level problem represents a decision-making process of system designers, in which possible highway alternatives are generated and evaluated. The lower-level problem represents highway users' route choice behavior under the designer's decision (i.e., the alternatives selected from the upper-level). The model optimizes the location of a new highway, including its intersection points with existing roads, and searches the best trade-off between the various highway cost components. An equilibrium traffic assignment is incorporated in the bi-level model framework to realistically reflect the traffic impact of the new highway in the alternative evaluation process.

In this chapter, we describe a bi-level HAO model, which finds highway alternatives that best improve the existing roadway system and optimizes their alignments based on geometric, cost, and operational considerations. The model is a mesoscopic highway route optimization model, which extends the microscopic GA-based HAO model discussed in Part II to include the user cost components (e.g., travel time and vehicle operation costs), by formulating it as a bi-level program. The model can now evaluate traffic operational impacts before and after a new highway implementation, while generating its possible alignments and evaluating detailed construction costs and environmental impacts. It is expected that highway planners and designers may greatly benefit from the proposed model, which offers well optimized candidate alternatives through the comprehensive bi-level optimization procedures rather than merely satisfactory alternatives obtained from the trial-and-error process in the traditional manual approach.

Despite the model's already demonstrated capabilities, it can still benefit from many technical and methodological improvements. Future work will seek to (i) improve model computational efficiency with a parallel genetic algorithm and distributed computing technique, (ii) improve the mechanism of highway alignment generation process with variable design speed, variable speed limit, and sensitive PI density to land-use and terrain complexity, (iii) develop an elastic demand function which can represent the induced demand by better level-of-service and by land-use change due to the new highway implementation, and (iv) introduce uncertainties (e.g., demand uncertainty and cost uncertainty) in the alternative evaluation process.

Chapter 11

Bi-level HAO Model Application Example

Chapter 11 presents an example study to demonstrate the performance of the bi-level HAO method. Figure 11.1 shows the land-use of the study area in which construction of a new highway is being considered for relieving the congestion in the existing highway system. A desktop PC (Pentium 4 CPU 3.2 GHz with 2GB RAM) is used for executing the bi-level alignment optimization model, and about six hours are taken to complete 300 generations of search. Land-use information and existing traffic condition of the study area are briefly described in the next.

11.1 Example Description

As described in Figure 11.1, *HW-1* is the only access control link connecting east-west traffic of the study area, and is operating at or near capacity during peak periods, causing severe traffic congestion. Furthermore, the number of trips within the study area is expected to increase in the near future due to new community developments. Thus, a local government is planning to construct a new highway for improving the level of service of the existing road, *HW-1* as well as for reducing users' travel time between traffic endpoints (i.e., Centroids in Figure 11.1).

Key input parameters and the base year traffic information used for this example are presented in Table 11.1. The baseline design standards of the new highway are a four-lane undivided highway with a 20 meter cross-section (3.6 meter for lanes and 2.8 meter for shoulders), a 90 km/h design speed, 6% maximum allowable gradient, 6% maximum superelevation. Totally $289(= 17 \times 17)$ Origin and Destination (O/D) trip pairs operate in the existing road network, and demand between east and west

Alternative	Total agency cost ($ Million)	Total user cost saving ($ Billion)	Penalty cost ($)	Length (feet)
Alt1	**39.0**	**-566.7**	**0.0**	**6,068**
Alt2	63.1	-571.2	0.0	5,445
Alt3	**39.4**	**-568.0**	**0.0**	**6,902**
Alt4	39.0	-267.2	0.0	3,076
Alt5	**40.2**	**-569.9**	**0.0**	**5,143**
Alt6	63.3	-569.0	0.0	4,260
Alt7	**39.6**	**-569.0**	**0.0**	**5,849**
Alt8	62.6	-293.1	0.0	4,244

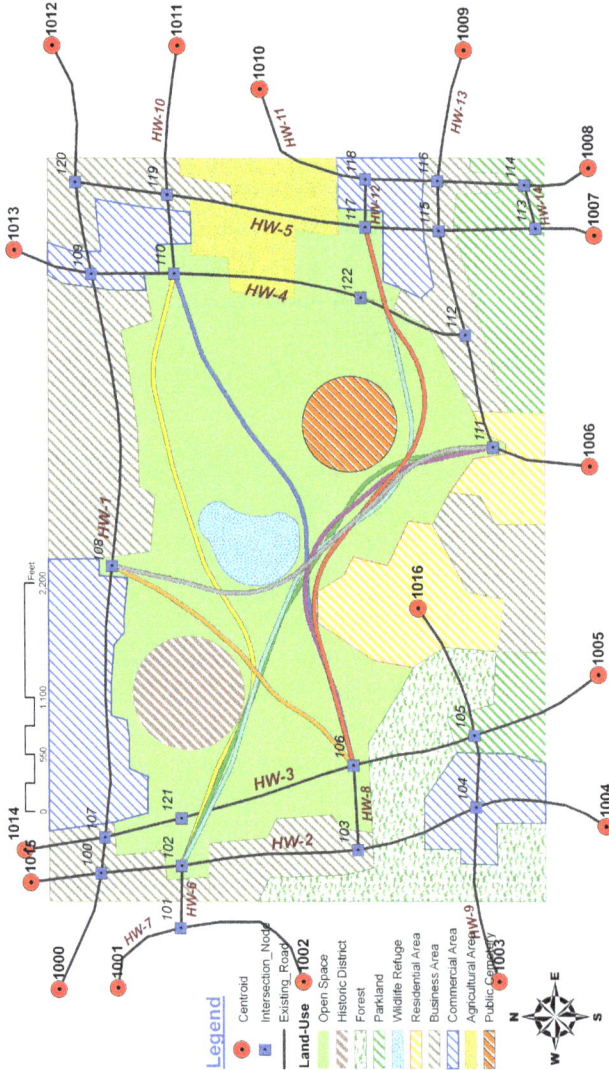

Figure 11.1   Selected optimized highway alternatives on an existing road network

Table 11.1 Key input parameters and a base-year O/D trip matrix

(a) Key input parameters

Input variable	Value
Road width	20 m
Lane width	3.6 m/lane
Paved shoulder width	2.8 m/shoulder
Design speed	90 kph
Maximum superelevation rate	6 %
Maximum allowable grade	6 %
Coefficient of side friction	0.12
Longitudinal friction coefficient	0.28
Fill/Cut slope	0.4, 0.5
Unit fill/cut cost	$3.3/m^3, $5.7/m^3
Earth shrinkage factor	0.9
Unit length-dependent cost	$328/m
Terrain height ranges	127 ~ 159 m
Unit land value in the study area	$0.1 ~ $452/m^2
Cross structure with existing road	Intersection
Annual traffic growth rate	3%
Annual interest rate	3%
Analysis period	5 years

(b) Base-year O/D trip matrix (veh/hr)

O/D	1000	1001	1002	1003	1004	1005	1006	1007	1008	1009	1010	1011	1012	1013	1014	1015	1016	Sum
1000	0	20	20	20	20	20	50	50	50	50	50	50	2000	50	20	20	20	2510
1001	20	0	20	20	20	20	50	50	50	50	50	50	50	50	20	20	20	560
1002	20	20	0	20	20	20	50	50	50	50	50	50	50	50	20	20	20	560
1003	20	20	20	0	20	20	50	50	50	50	50	50	50	50	20	20	20	560
1004	20	20	20	20	0	20	50	50	50	50	50	50	50	50	20	20	20	560
1005	20	20	20	20	20	0	50	50	50	50	50	50	50	50	20	20	20	560
1006	50	50	50	50	50	50	0	20	20	20	20	20	20	20	50	50	50	590
1007	50	50	50	50	50	50	20	0	20	20	20	20	20	20	50	50	50	590
1008	50	50	50	50	50	50	20	20	0	20	20	20	20	20	50	50	50	590
1009	50	50	50	50	50	50	20	20	20	0	20	20	20	20	50	50	50	590
1010	50	50	50	50	50	50	20	20	20	20	0	20	20	20	50	50	50	590
1011	50	50	50	50	50	50	20	20	20	20	20	0	20	20	50	50	50	590
1012	2000	50	50	50	50	50	20	20	20	20	20	20	0	20	50	50	50	2540
1013	50	50	50	50	50	50	20	20	20	20	20	20	20	0	50	50	50	590
1014	20	20	20	20	20	20	50	50	50	50	50	50	50	50	0	20	20	560
1015	20	20	20	20	20	20	50	50	50	50	50	50	50	50	20	0	20	560
1016	20	20	20	20	20	20	50	50	50	50	50	50	50	50	20	20	0	560
Sum	2510	560	560	560	560	560	590	590	590	590	590	590	2540	590	560	560	560	13660

traffic endpoints is much higher than north-south traffic demand. The annual traffic growth rate is assumed to be 3%.

The new highway should be constructed in an environmentally responsible way since various socio-economic and environmentally sensitive areas (e.g., residential area, commercial area, historic district, and wildlife refuge) are mixed in the study area. With all these considerations, the objective of the local government to the new highway project can be as follows:

- The new highway should connect the existing and planned development areas and must be an economical path that minimizes the highway agency cost.
- It should relieve congestion on existing highways (i.e., minimize total user cost).
- It should minimize environmental impact.
- It should minimize socio-economic impact.

11.2 Optimized Alternatives

Eight highway alternatives are selected after the model completes the optimization process. Each of them is the best-obtained solution for a given pair of start and end points. Figure 11.1 shows horizontal profiles of the selected highway alternatives, including their total construction costs and user cost savings. As shown in the figure, all of them fully avoid the restricted areas (e.g., wildlife refuge, residential area, and public cemetery) located in the middle of study area, and thus do not have any environmental and socio-economic impacts (i.e., no penalty cost). Among the alternatives, **Alt-2**, **Alt-6**, and **Alt-8** would be ruled out by highway designers if the project budget is limited to $45 million. **Alt-8** is the worst option among the selected alternatives since it requires almost the highest agency cost and saves less user cost compared to other alternatives. **Alt-4** requires the least agency cost, and thus it would be the best alternative if the user cost is not included in the evaluation criteria. However, it is also ruled out since it does not significantly improve the existing traffic operation (i.e., the least user cost saving). Thus, **Alt-1**, **Alt-3**, **Alt-5**, and **Alt-7** are preferable options since their agency costs are within the project budget and their user costs are significantly lower than for the other alternatives.

Table 11.2 shows the equilibrium link flows operating on the existing and new highways before and after the new highways' implementation. The results demonstrate that the equilibrium link flows can be greatly affected by the highway alignment, particularly in terms of distance and intersection points (i.e., whether it connects within the network). The table also shows that **Alt-1** and **Alt-3** should be excluded from the preferable alternative set since some existing highways (e.g., *HW-3*, *HW-4*, and *HW-5*) may operate slightly over the capacity (only according to the Frank-Wolfe algorithm) if these alternatives are implemented.

Assuming **Alt-5** is the best alternative (since it requires the least objective function value, which is the sum the total agency cost and the user cost), Table 11.3 summarizes the estimated traffic improvements on major highways after the new highway addition. Traffic condition of the *HW-1* can be significantly improved with 25% traffic reduction after the new highway implementation. This indicates that a large fraction of traffic from *HW-1* transfers to the new highway to save travel times. Traffic on *HW-3* and *HW-4* can also be improved (with 39 and 26% reduction, respectively) due to the new highway construction.

Because the objective function of the optimization model consists of the eight cost components (earthwork, right-of-way, length-dependent, structure, maintenance, travel time, vehicle operation, and penalty costs), it attempts to find the best trade-off between the various cost components and finally obtains the minimum total. An analysis is performed to investigate the fraction of the various cost items in the objective function value of the optimized highway alignment (here **Alt-5**). A detailed breakdown of its objective function value is shown in Table 11.4. As shown in the table, the earthwork cost and structure cost make up the first and second highest fractions of the total agency cost, dominating the other agency cost items. From the user cost point of view, the new highway significantly reduces users' travel time cost and vehicle operation cost, which dominate all the agency cost items in the objective function. No penalty cost is involved in the new highway construction. Note that the negative values of the user cost items indicate that the user costs estimated before the system improvement are greater than those after the new highway implementation.

Table 11.2 Equilibrium link flows on major highways before/after new highway implementation

Alternative	Distance (feet)	Intersection Point (node #)	Link flow (vph)	Existing highway (capacity)				
				HW-1 (8800vph)	HW-2 (3600vph)	HW-3 (4000vph)	HW-4 (4400vph)	HW-5 (3600vph)
No Build				8628 (0.98)	934 (0.26)	3584 (0.90)	4454 (1.01)	735 (0.20)
Build								
Alt-1	6068	102, 110	7950 (0.99)	2655 (0.30)	1127 (0.31)	4957 (1.24)	4554 (1.04)	715 (0.20)
Alt-2	5445	102, 111	7888 (0.99)	5505 (0.63)	1127 (0.31)	3324 (0.83)	1654 (0.38)	510 (0.14)
Alt-3	6902	102, 117	6894 (0.86)	3690 (0.42)	1127 (0.31)	4957 (1.24)	1100 (0.25)	3835 (1.07)
Alt-4	3076	106, 108	3200 (0.40)	7908 (0.90)	774 (0.22)	1717 (0.43)	4454 (1.01)	748 (0.21)
Alt-5	5143	106, 110	3572 (0.45)	6497 (0.74)	1326 (0.37)	2198 (0.55)	3300 (0.75)	1069 (0.30)
Alt-6	4260	106, 111	4500 (0.56)	6458 (0.73)	774 (0.22)	4184 (1.05)	554 (0.13)	485 (0.13)
Alt-7	5849	106, 117	4190 (0.52)	1668 (0.19)	3114 (0.87)	344 (0.09)	1534 (0.35)	1210 (0.34)
Alt-8	4244	108, 111	4500 (0.56)	6825 (0.78)	934 (0.26)	3584 (0.90)	1457 (0.33)	485 (0.13)

Note: Measures within parentheses represent V/C Ratios.

Table 11.3 Traffic improvement on major highways after implementation of Alt-5

Road (ID)	Capacity(vph)	V/C Ratio No build	Build	Traffic reduction (%)
Alt-5	8,000	—	0.45	—
HW-1	8,800	0.98	0.74	24.49
HW-2	3,600	0.26	0.37	—
HW-3	4,000	0.90	0.55	38.89
HW-4	4,400	1.01	0.75	25.74
HW-5	3,600	0.21	0.30	—

Table 11.4 Total cost breakdown of the optimized alignment, Alt-5

Type of cost	Million ($)	Fraction (%)
Total agency cost	40.2	100.00
Earthwork	20.0	49.68
Length-dependent	1.0	2.56
Right-of-way	0.5	1.16
Structures	18.6	46.38
Maintenance	0.1	0.22
Total user cost	−569.9	100.00
Travel time	−427.7	75.05
Vehicle operation	−142.2	24.95
Penalty cost	0.0	0.00
Total objective function value	−529.7	100.0

11.3 Summary

The performance of the bi-level HAO model is demonstrated with a hypothetical example. The results show that the model can find optimized solutions within reasonable computation times, and that locations of new highways are sensitive to traffic distributed to the road network besides their construction costs. This confirms that all relevant highway cost components should be simultaneously evaluated for an effective highway alignment optimization although many highway agencies tend to ignore the user cost items in the planning phase of new highways.

The applicability of the bi-level HAO model may vary depending on the characteristics and scopes of given highway projects. For a small and short highway project (e.g., less than one mile long), which only focuses on minimizing the construction cost and right-of-way impact, it may suffice to use the alignment optimization which is the upper-level method in the proposed model. However, a large scale highway project which may significantly affect traffic demand pattern (such as a highway strategic master-plan which identifies existing and future needs of new highways in accordance with growing traffic demand) may greatly benefit from fully using the bi-level method.

Part IV

Highway Alignment Optimization Model Applications and Extensions

Chapter 12

HAO Model Application in Maryland Brookeville Bypass Project

12.1 Project Description

The Maryland State Highway Administration (MDSHA) has been working on the MD 97 Brookeville Bypass project in Montgomery County, Maryland. This area is listed on the National Register of Historic Places as a historic district, and is located approximately ten miles south of I-70 and three miles north of MD 108. The project objectives are to divert the increasing traffic volumes from the town of Brookeville by constructing a new bypass route to improve traffic operation and safety on existing MD 97, while preserving the historic character of the town. The alignment optimization model is tested in the real highway construction project to assist the local government in finding the best alternatives while considering various issues arising in the project.

Through this case study we (i) demonstrate the applicability and usability of the model to a real highway project with due consideration to issues arising in real-world applications and (ii) analyze the sensitivity of solution alignments to various user-specified input variables (such as the number of points of intersection (PI's), composition of the model objective function, and design speed); in addition, (iii) goodness of the solutions found from the model is statistically evaluated. To ensure comparability with the normal evaluation criteria typically used by the highway agencies, such as those used by the MDSHA, the user cost which consists of travel time cost, vehicle operating cost, and the accident cost is suppressed from the model objective function. Thus, the objective function used in applying the model to the Brookeville project is $C_{T\_Agency} = C_L + C_R + C_E + C_S + C_M$.

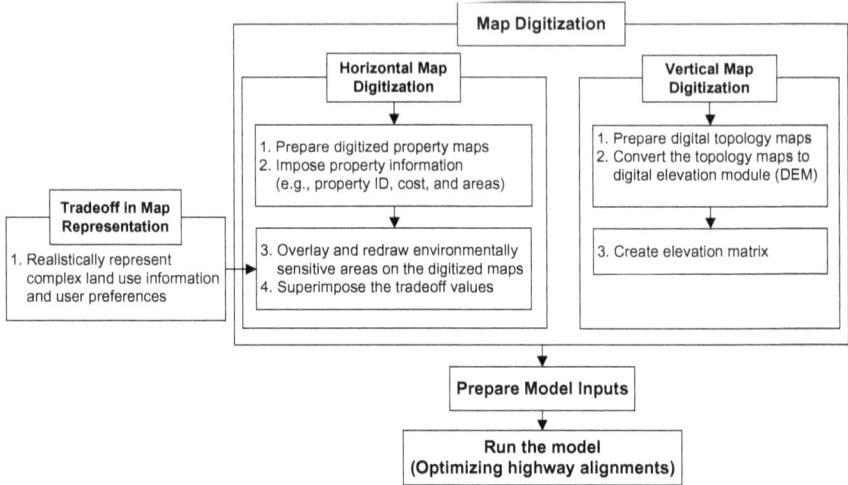

Figure 12.1 Model application procedures for the Brookeville Bypass project

12.2 Data and Application Procedure

Three major data preprocessing tasks are performed before optimizing highway alignments with the model; (i) horizontal map digitization, (ii) vertical map digitization, and (iii) tradeoff in map representation. Figure 12.1 presents the application procedure of the model to the Brookeville Bypass project. Such data preprocessing is necessary for the HAO model to directly access customized GIS maps during the alignment optimization process. For preparation of the GIS inputs we used as-built plans of existing roads, a digital elevation raster dataset (or a study area elevation map in a CADD file format), and Maryland's GIS database (called MdProperty View), which includes various land-uses, land values, and information on natural resource and man-made features in the project area.

12.2.1 Horizontal Map Digitization

For horizontal map digitization, MicroStation base-maps which store boundaries of environmentally sensitive areas, such as wetlands, floodplains, and historic resources are used to digitize properties in the study area of Brookeville. In this step, each property is regarded as a polygon, which can retain property information as its attributes. The purpose of horizontal map digitization is to reflect complex land-uses in the study area on the GIS digitized map, and eventually to use it for evaluating the detailed alignment right-of-way cost and environmental impacts during the optimization process. The information assigned on the map includes parcel ID number, perimeter, unit cost, and area of each property.

As shown in Figure 12.2, the study area combines various types of natural and cultural land-use patterns. There are 10 different types of land-use characteristics in the study area: structures (houses and other facilities), wetlands, residential areas, historic places, streams, park with historic district, parklands, floodplains, existing roads, and other properties. Note that such a map superimposition is pre-processed with the IDPM, and is essential for applying the feasible gates (FG) methods which is designed for representing the user preferences in the alignment optimization process effectively. Through preprocessing, the model users can define feasible bounds of solution alignments generated by the model (see Chapter 8

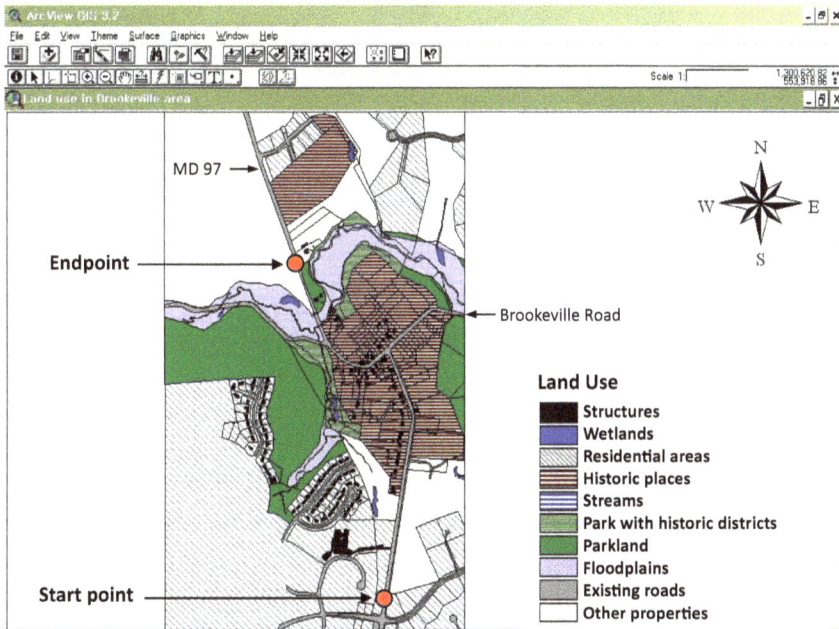

Figure 12.2 Land use of the study area for Brookeville Bypass project

for the details of the FG methods). The study area comprises about 650 geographic entities (including land, structures, road, etc.) with given start and end points of the proposed alignment. The 690 acres ($2.792\,\text{km}^2$) of the search space includes 203.3 acres ($0.823\,\text{km}^2$) of primarily residential areas, 73.4 acres ($0.297\,\text{km}^2$) of historic sites, 67.5 acres ($0.273\,\text{km}^2$) of parkland, and 30.9 acres ($0.125\,\text{km}^2$) of floodplains.

12.2.2 Vertical Map Digitization

In the model, the alignment earthwork cost is calculated based on a ground elevation, whose preparation for the study area is required. To do this, we use a Microstation contour map for the study area, and convert it to a Digital Elevation Model (DEM) that provides elevations with a grid base as shown in Figure 12.3. The study area is divided into evenly spaced grids of 40×40 feet (12×12 meters). Finer grids may be selected for precise earthwork calculation if desired. The elevation range in the Brookeville area is 328 to 508 feet (100 to 155 meters). The darker areas represent higher elevations. Floodplains and parklands exist in low elevation areas while the historic places are located at relatively high elevations (refer to Figures 12.2 and 12.3).

12.2.3 Tradeoffs in Map Representation for Environmental Issues

When considering roadway construction in a given project area, various geographically sensitive regions (such as historic sites, creeks, public facilities, etc.) may be encountered. These control areas should be avoided by the proposed alignment, whose impact on these regions should be minimized as much as possible. Based on a previous Brookeville study by MDSHA (2013), we recognize residential properties, the Longwood Community center, historic districts, and wetlands as primary environmentally sensitive areas that should be avoided by the new alignments if at all possible (i.e., those are untouchable areas). In addition, parklands, floodplains, and streams, which are located between the given start and end points and cannot avoid being taken by the proposed alignment, are considered secondary environmentally sensitive areas.

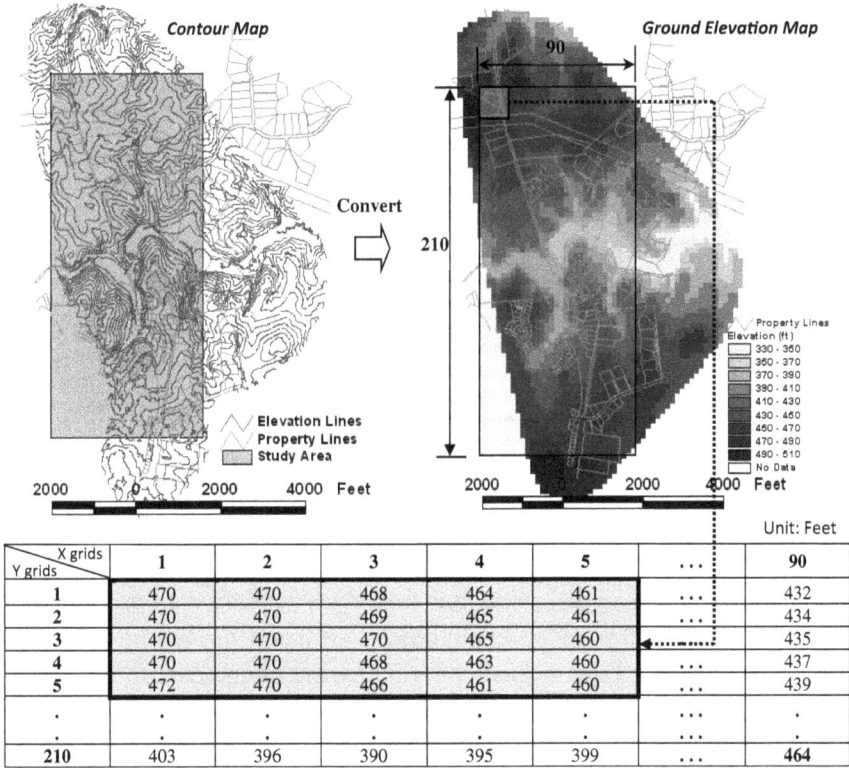

Figure 12.3 Ground elevation of the Brookeville Project area

To realistically represent such control areas in the model application, we divide them into two categories based on their land-use characteristics as shown in Table 12.1: Type 1 areas that the proposed roadway alternatives can avoid, and Type 2 areas that the proposed alternatives cannot avoid. Type 1 areas include wetlands, historic places, residential areas, Community Center, and other structures. Type 2 areas consist of streams, parklands, and floodplains which are unavoidably affected by the alignments. To properly reflect these relevant environmental issues in the GIS map representation, tradeoff values with respect to the different land use types must be carefully determined based on their relative importance, since these values can significantly affect the resulting alignment. Thus, the maximum allowable areas affected by the new alignments (denoted as *MaxA*) should be much

Table 12.1 Spatial control areas in the Brookeville Bypass project

Type	Control areas	Characteristics	*MaxA*
Type1	• Wetlands, historic places, residential properties, site of community center, structures (houses, public facilities, etc.)	• The control areas that the proposed alignment can avoid	0
Type2	• Streams, floodplains, parklands	• The control areas that the proposed alignment cannot avoid	User-specifiable

stricter for Type 1 areas than for Type 2 areas; recall that Type 1 areas have primary (i.e., stronger) environmental regions to be avoided by the alignments whereas Type 2 areas contain only secondary regions. Note that this idea seeks to eliminate the alignments' impacts on Type 1 areas and minimize those on Type 2 areas, by guiding the alignments to take other properties, which have no restrictions. For this purpose, we discriminate between Type 1 and Type 2 areas by assigning different values of *MaxA*. For the environmentally sensitive regions classified as Type 1 areas, their *MaxA* are set to be 0 (which means Type 1 areas are not allowed to be affected by the new alignments), while the *MaxA* of control areas defined as Type 2 can be interactively specified by the model users based on their relative importance.

12.2.4   Description of Model Inputs and Outputs

For optimizing highway alignments with the model, some input variables must be pre-specified. These are, for instance, road width, design speed, and maximum vertical gradient of the proposed alignments. Since the optimized alignment varies depending on these inputs, users should carefully determine the input variable values. The start and end points of the new alignments are assumed to be known in this case study. They are located on the south and north sections of MD 97 in Brookeville, respectively, as shown in Figure 12.2. The Euclidean distance between the start and end points is

Brookeville bypass cross section

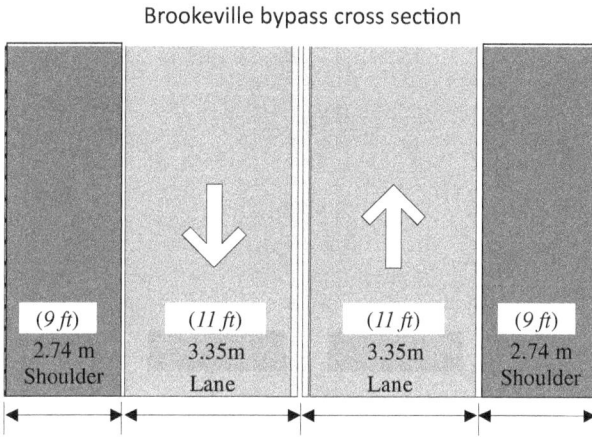

Figure 12.4 Cross section of a new highway alignment for Brookeville Bypass project

about 0.76 mile (1.22 km). The design speed was initially set at 50 mph (80kph). The distances between station points (i.e., cross-section spacing), which are used as earthwork computation unit in the model formulation, are assumed to be 50 feet (15 meters).

The cross-section of the proposed alignment is assumed to represent a two-lane road with a 40 foot width (11 feet for lanes and 9 feet for shoulders as shown in Figure 12.4). In addition, grade separation is assumed to be the only crossing type of the new highway with the existing Brookeville Road. Various user-specifiable input variables required in the highway alignment optimization process are described on the left side of Table 12.2. The values on the right side of the table are used for the model application to the Brookeville example. The unit road construction costs, such as unit cut and fill costs and length-dependent costs are user-specifiable. The total cost of a solution alignment is computed based on the pre-specified unit costs.

Note that detailed results for the optimized alignments, such as total cost breakdown, earthwork cost per station, and coordinates of all evaluated alignments, are provided as the model outputs. These results are automatically restored in different files during program runs. In addition, alignments' impacts to the environmentally sensitive areas can also be summarized using the GIS module embedded in the optimization model.

Table 12.2 Baseline inputs used in the model application to Brookeville Bypass project

Input variables	Value
No. of intersection points (PI's)	4~7
Road width	40 foot, 2-lane road (11"lane, 9"shoulder)
Design speed	50 mph (80 kph)
Maximum superelevation	0.06
Maximum allowable grade	5 %
Coefficient of side friction	0.16
Longitudinal friction coefficient	0.28
Distance between station points	50 feet (15 meters)
Fill slope	0.4
Cut slope	0.5
Earth shrinkage factor	0.9
Unit cut cost	35 $/yard^3 (45.5$/m^3)
Unit fill cost	20 $/yard^3 (26$/m^3)
Cost of moving earth from a borrow pit	2 $/yard^3 (2.6 $/m^3)
Cost of moving earth to a fill	3 $/yard^3 (3.9 $/m^3)
Unit length-dependent cost*	400 $/feet (656 $/meter)
Crossing type with the existing road	Grade separation
Terrain height ranges	328~508 feet (100~155 meter)
Unit land value in the study area	0~14 $/ft^2 (0~151 $/m^2)

Unit length-dependent cost mainly consists of unit pavement cost and sub and super structure (e.g. barrier and median) costs on the road.

12.3 Optimization Results

12.3.1 Optimized Alignments with Different Numbers of PI's

Optimizing (roughly) the number of PI's is quite desirable in a model application to real highway construction projects because more PI's may be preferable for alignments in complex and high density areas (such as urban areas), while fewer PI's may suffice for projects in the rural study areas. For instance, applying the lower number of PI's (e.g., 2 or 3 PI's) to the Brookeville example may not be sufficient to keep solution alignments away from the complex control areas. It should also be noted that the solution quality (such as alignment's impact on the environmentally sensitive areas

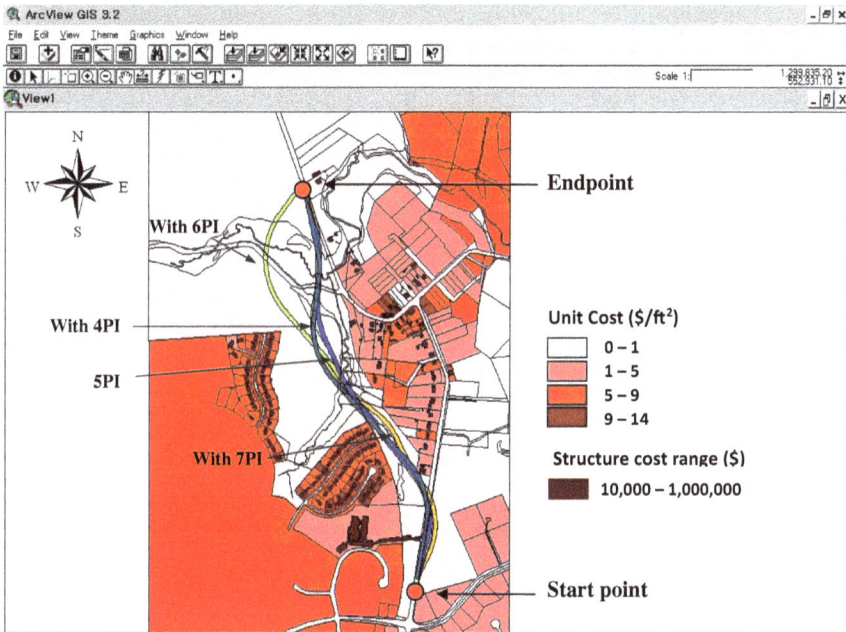

Figure 12.5 Optimized horizontal alignments with different numbers of PI's for Brookeville Bypass project

Optimized alignment	# of PI's	Total agency costs ($)	Environmental impact			Residential relocation (No.)	Length (ft)	Computation time (hr)
			The control area taken by alignments (ft²)					
			Type 1	Type 2	Sum			
A	4	4,847,128	458.3	70,674.2	71,132.6	0	4,359.9	4.41
B	5	4,328,432	0.0	63,030.4	63,030.4	0	4,302.0	4.68
C	6	5,655,707	0.0	82,017.4	82,017.4	0	4,607.3	4.95
D	7	4,919,403	0.0	64,489.3	64,489.3	0	4,422.9	5.01

and right-of-way cost) and computation efficiency of the model may vary depending on the number of PI's.

The optimized solution alignments found with the inputs presented in Table 12.2 are shown in Figures 12.5 and 12.6. To explore the preferable number of PI's, the model was run four times with four to seven PI's. As shown in Figure 12.5, horizontal profiles of the optimized alignments A, B, C, D have 4, 5, 6, and 7 PI's, respectively. Vertical profiles of those alignments are presented in Figure 12.6. Note that more than 8 PI's were not

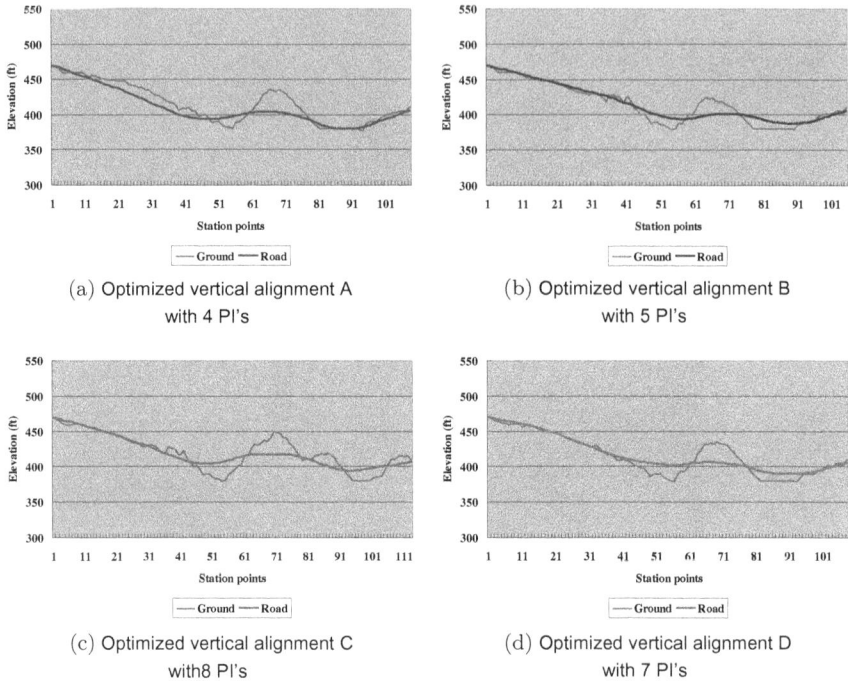

(a) Optimized vertical alignment A
with 4 PI's

(b) Optimized vertical alignment B
with 5 PI's

(c) Optimized vertical alignment C
with 8 PI's

(d) Optimized vertical alignment D
with 7 PI's

Figure 12.6 Optimized vertical alignments with different numbers of PI's for Brookeville Bypass project

considered in this case study since they might create too many horizontal curves and increase the model's computation time. For each of the four cases, the model searched over 300 generations, thereby evaluating 6,500 alignments. A desktop PC Pentium IV 3.2 GHZ with 2 GB RAM was used to run the model. It took considerable time (about 4.5 to 6.5 hours) to run through 300 generations because the Brookeville study area is quite complex and has many properties (about 650 geographical entities).

As shown in Figures 12.5 and 12.6, the horizontal and vertical profiles of the four optimized alignments seem very similar. They have similar rights-of-way and alignment lengths; in addition, none of the four alternatives require any residential relocation. However, it should be noted that detailed model outputs, such as total agency costs and environmental impacts of those alignments are quite different. Among the four alternatives, the lowest agency cost is $ 4,328,432 and the highest cost $ 5,655,707.

In terms of environmental impact, the sensitive areas taken by the optimized alignment B (63,030.4 ft^2 for total) are the lowest, although the differences among the four alternatives are not great (see Figure 12.5). For Type 1 areas, which were previously defined as primary sensitive regions, optimized alignment A with four PI's affects relatively large amounts of Type 1 areas compared to those of the other three alternatives. Alignment A affects 458.3 ft^2 of Type 1 areas (306 ft^2 for residential area and 152.3 ft^2 for Longwood Community Center); the other three optimized alignments do not affect Type 1 areas. A more detailed environmental impact summary for the four alternatives is presented in Table 12.3.

In terms of computation efficiency (refer to Figure 12.5), the model's computation time increases slightly when the number of PI's increases from four to seven. It seems that computation time is not greatly affected by the number of PI's. However, we note that computation time still increases with the number of PI's since additional PI's generate additional horizontal and vertical curved sections. For instance, a model application with 20 input PI's for the same example project requires over 10 hours of computations.

It should be noted that the total agency cost estimated from the model (see Table 12.3) is underestimated. This cost mainly consists of length-dependent, right-of-way, earthwork cost, structure cost, and maintenance cost; i.e., other agency costs required in the road construction (such as drainage landscape architecture cost, traffic signal strain poles cost, etc.) and contingency cost are not included. Readers may refer to Table 7.5 in Chapter 7 to see the detailed breakdown of the total agency cost of the optimized alignment found with five PI's (i.e., alternative B). In addition, Figure 8.6 presented in Chapter 8 shows the changes in objective function values over successive generations for that case.

Figure 12.7 shows one of the most favored alternatives obtained manually by the MDSHA. The agency has obtained the alternative through a repetitive time-consuming process, while the proposed model can allow a quick alignment optimization process (within several hours) with remarkable precision. A comparison of the MDSHA's alternative with the one optimized by our model indicates that the configurations of the two alternatives seem similar; however, the total costs estimated and other important measures of effectiveness (MOEs) (e.g, total length and affected historic districts and wetlands) used for evaluating the alternatives are different. This

Table 12.3 Environmental impact summary for optimized alignments A to D

Optimized alignments		A	B	C	D
Number of PI's		4	5	6	7
Total agency cost (million $)		4.86	4.33	5.66	4.92
Alignment length (feet)		4,359.9	4,302.0	4,607.3	4,422.9
Socio-economic resources	Residential area affected (ft^2)	305.96	0	0	0
	Residential relocations (no.)	0	0	0	0
	Community center affected (ft^2)	152.38	0	0	0
	Historic places affected (ft^2)	0	0	0	0
	County reserved areas affected (ft^2)	41,896.1	45,295.9	45,286.0	45,260.0
	Existing roads affected (ft^2)	39,152.1	29,609.1	17,037.6	25,227.4
Natural resources	Wetlands affected (ft^2)	0	0	0	0
	Floodplains affected (ft^2)	23,259.8	17,260.3	16,689.7	14,883.5
	Streams affected (ft^2)	690.5	777.6	634.9	610.7
	Parkland affected (ft^2)	46,723.8	44,992.5	64,692.7	48,995.2

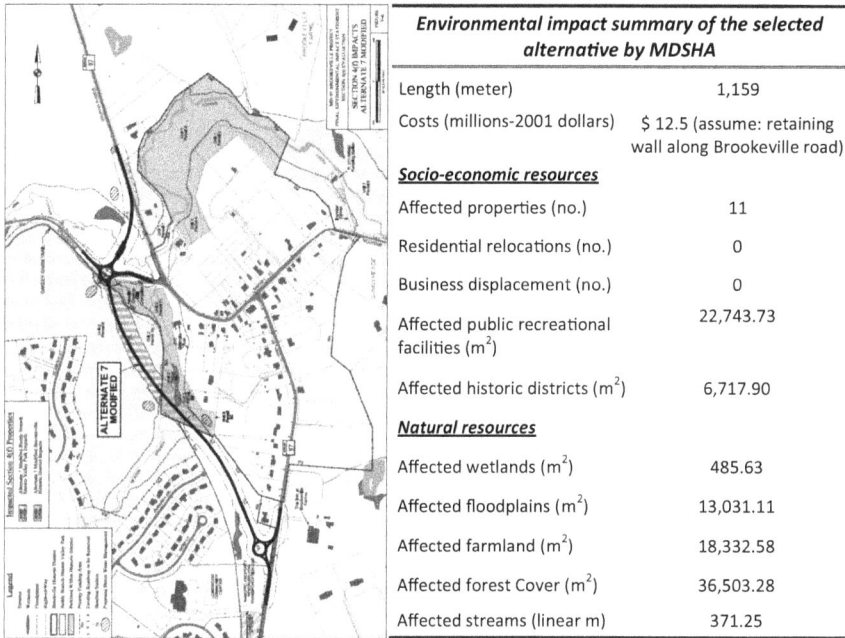

Environmental impact summary of the selected alternative by MDSHA	
Length (meter)	1,159
Costs (millions-2001 dollars)	$ 12.5 (assume: retaining wall along Brookeville road)
Socio-economic resources	
Affected properties (no.)	11
Residential relocations (no.)	0
Business displacement (no.)	0
Affected public recreational facilities (m^2)	22,743.73
Affected historic districts (m^2)	6,717.90
Natural resources	
Affected wetlands (m^2)	485.63
Affected floodplains (m^2)	13,031.11
Affected farmland (m^2)	18,332.58
Affected forest Cover (m^2)	36,503.28
Affected streams (linear m)	371.25

Figure 12.7 MDSHA's selected alternative for the Brookeville Bypass project

is largely due to the resource limitations imposed by the manual method on the search for alternatives, sensitivity analysis, and trade-off analysis. It should also be noted that the total cost of the optimized alignment is somewhat underestimated since it does not include several miscellaneous highway costs (e.g., drainage and contingency costs) which are included in the manual solution.

12.3.2 Goodness Test

Although the solution alignments found with the HAO model seems to be reasonable, we aim to evaluate how good the solutions are. For this purpose, an experiment is designed to statistically test the goodness of the solutions found by the model. Table 12.4 describes three different scenarios of the experiment. In this experiment, the 1st scenario is initiated by randomly generating solutions to the problem (i.e., sample solutions are generated from a random search process). The 2nd scenario is a random search process with human judgments. This scenario is designed for representing a

Table 12.4 Three test scenarios for assessing goodness of solutions found from the HAO model

Scenario	Description	Search option	Iteration
1	Random generation of PI's in the entire study area	Random search	15,000
2	Random generation of PI's within the user-defined feasible bounds	Random search with FG, and P&R methods	15,000
3	Evolutionary search of PI's within the user-defined feasible bounds	GAs-based search with FG and P&R methods	9,050*

About 9,050 alignments are generated for 300 generations.

path selection process of a new highway conducted in an actual road construction project. For this purpose, it is assumed that spatial information about no-go areas (i.e., untouchable areas) that new alignments must avoid is already known and that all generated solutions should meet given design constraints. Such a scenario is implemented by applying (i) the feasible gate (FG) method (for identifying the user-defined alignment feasible boundaries) and (ii) prescreening and repairing (P&R) method (for maintaining the required design specifications) to a random search process. In the last (3rd) scenario the proposed optimization model is used to search for alignments (i.e., search with the customized GAs integrated with the FG and P&R methods). Note that the first two scenarios have no learning procedure during the search process; however, the 3rd scenario is an adaptive search based on the principles of natural evolution and survival of the fittest.

Fifteen thousand sample solutions are created from each of the 1st and 2nd scenarios. Note that these samples are created in such a way that the solutions are representative and independent of each other. Figure 12.8 shows distribution diagrams of the sample solutions generated from the 1st and 2nd scenarios. In order to ascertain the goodness of the optimized solution found from the 3rd scenario (i.e., optimized alignment B), its relative position is also indicated on that figure.

As shown in Figure 12.8, it is observed that the objective function values of the solution alignments found from the two random search processes (the 1st and 2nd scenarios) are very widely distributed (1st scenario: 18.2 $\sim$ 9, 945.0 millions; 2nd scenario: 11.8 $\sim$ 2, 802.5 millions) and there are two

(a) Distribution of objective function values

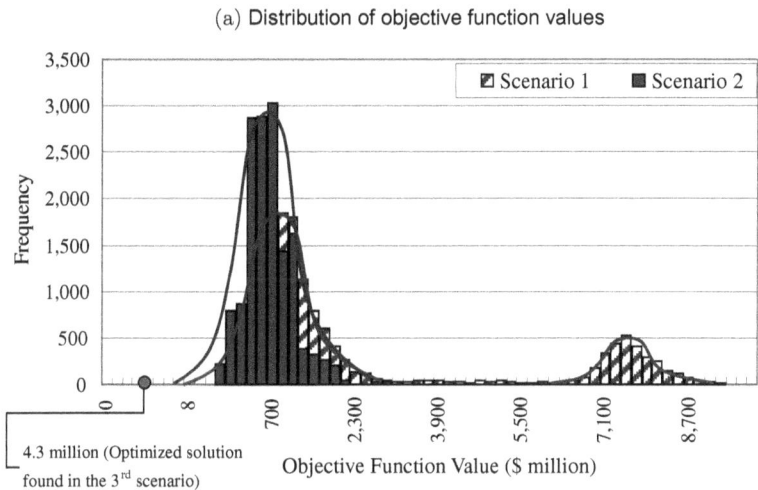

4.3 million (Optimized solution found in the 3rd scenario)

Objective Function Value ($ million)

(b) Descriptive statistics of objective function values

Scenario	Min	Max	Mean	Median	Standard Deviation
1	18.2	9,945.0	2,355.7	1041.7	2,699.4
2	11.8	2,802.5	547.9	493.8	412.8
3	4.3*				

* Objective function value (i.e., total agency cost) of the optimized alignment B

Figure 12.8 Experiment testing the goodness of optimized solutions

distinct high frequency ranges in the solution distribution of the 1st scenario case and one in the 2nd scenario case. Our interpretation for the 1st scenario result is that very high cost properties (such as house and building structures) are spatially distributed (scattered) only in specific regions of the study area as shown in Figure 12.5, and thus some fractions of alignments generated from the random search processes (i.e., the 1st scenario) can possibly cross those properties but some may not. That is why there are two distinct high frequency ranges in the distribution of the 1st scenario.

On the other hand, rights-of-way of alignments resulting from the 2nd scenario are mainly placed on the other properties (e.g., farm and park-lands) whose costs are relatively low and range widely, while avoiding the high cost structures. This occurs because the 2nd scenario is designed to perform a random search process within a feasible boundary that represents user preferences (see Figure 7.8 in section 7.4 for the user-specified feasible boundary). However, it should be noted that their objective function values are still higher than those from the optimization model since other cost components included in the objective function (such as earthwork and length-dependent costs) increase the total objective function values without being optimized.

Figure 12.8 also shows that the objective function value (about 4.3 million) of the optimized alignment from the model is considerably lower than the lower bounds (18.2 and 11.8 million) of the sample distributions from the two random search scenarios. This indicates that it dominates all possible solutions in the sample distribution. Thus, the solution found by the model is remarkably good when compared with other possible solutions to the problem.

12.4   Sensitivity of Optimized Alignments to Other Major Input Parameters

Beyond the number of PI's, the sensitivity to other major input parameters of the alignment optimization model (such as components of objective function, design speed, elevation grid size, and cross-section spacing) is also examined in this section. To check the influence of such factors on the solution quality, the input data values implemented for optimized alignment B (see inputs in Table 12.2 with five PI's) are used as the default values since it seems the preferable one according to the results presented in Table 12.3; its initial construction cost is the lowest and it does not affect any spatially sensitive area.

12.4.1   Sensitivity to the Model's Objective Function

The sensitivity of optimized alignments to various cost components associated with alignment construction is tested here. This analysis is intended to show the effect of various model objectives so as to emphasize that all the alignment-sensitive costs should be considered and precisely formulated for a good highway optimization model. Three different scenarios are designed to show how each cost component affects the resulting alignments. Note that all the solution alignments found in this sensitivity analysis adopt the same input parameters besides the cost components composing of the objective function as follows:

- Case 1: $C_{T\_Agency} = C_L + C_S + C_M$ (i.e., the sum of length-dependent, structure, and maintenance costs for the objective function)
- Case 2: $C_{T\_Agency} = C_L + C_S + C_M + C_R$ (i.e., the right-of-way cost is added to Case 1)
- Case 3: $C_{T\_Agency} = C_L + C_S + C_M + C_R + C_E$ (i.e., the earthwork cost is added to Case 2)

As shown in Figure 12.9(a1), the solution alignment being optimized only with $C_L + C_S + C_M$ is a straight line horizontally and affects many high-cost and environmentally sensitive areas (e.g., residential and historic areas). In addition, Figure 12.9(b1) shows that its vertical profile (i.e., road elevation) is hugely different from the corresponding ground elevation profile and obviously not optimized. Such results occur because the solution alignment is optimized with the objective function that does not represent the complexity of land use system and topography of the study area. Note that the objective function of this case does not include the right-of-way cost, environmental impacts on the sensitive areas, and earthwork cost.

Figure 12.9(a2) shows the horizontal profile of the optimized alignment found with the four cost components ($C_L + C_S + C_M + C_R$) of the objective function (i.e., Case 2). As shown in the figure, this alignment hardly affects the expensive land areas and is relatively circuitous in avoiding the environmentally sensitive areas. However, its vertical alignment is still not optimized (i.e., it still has a huge difference with the ground elevation) since the model objective function of Case 2 does not consider the earthwork cost component [see Figure 12.9(b2)].

The horizontal and vertical profiles of the optimized alignment resulting when we consider all the five major costs ($C_L + C_S + C_M + C_R + C_E$) are presented in Figures 12.9(a3) and 12.9(b3), respectively. Although the horizontal profile of this resulting alignment (Case 3) is similar with that of Case 2, its vertical alignment is quite different. As shown in Figure 12.9(b3) its vertical profile closely follows the ground elevation. This is because horizontal and vertical alignments are optimized jointly while minimizing its earthwork cost as well as the other four cost components. Note that the structure cost (C_S) and maintenance cost (C_M), although also dominating

(a1) Optimized alignment
with $C_T=C_L+C_S+C_M$

(Case1)

(a2) Optimized alignment
with $C_T=C_L+C_S+C_M+C_R$

(Case2)

(a3) Optimized alignment
with $C_T=C_L+C_S+C_M+C_R+C_E$

(Case3)

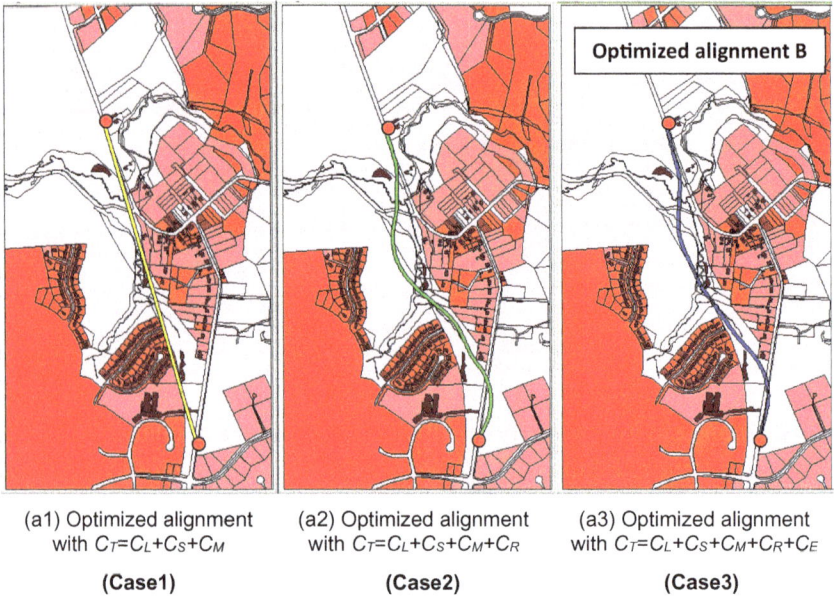

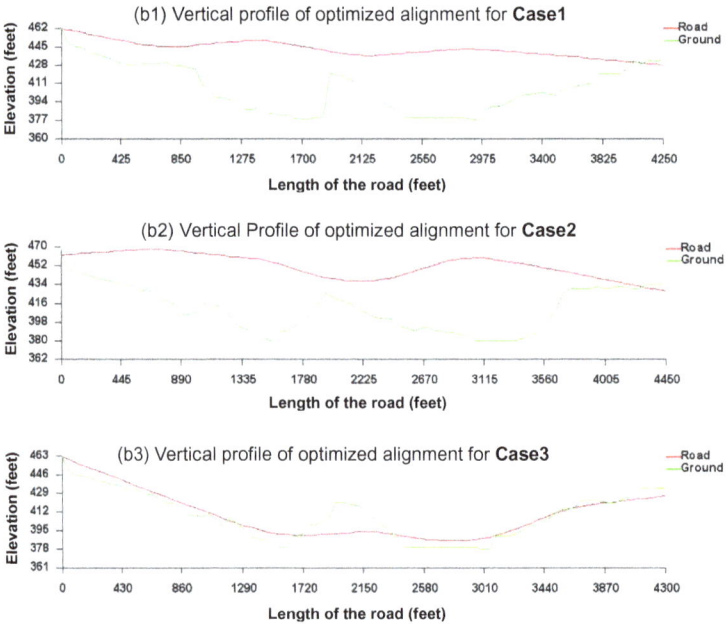

Figure 12.9   Sensitivity of optimized alignments to objective function

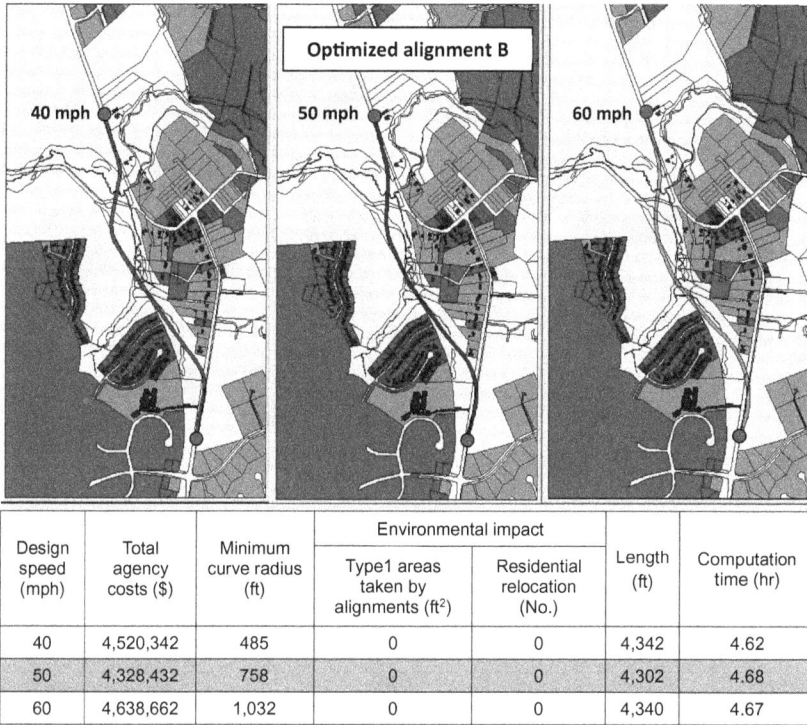

Design speed (mph)	Total agency costs ($)	Minimum curve radius (ft)	Environmental impact		Length (ft)	Computation time (hr)
			Type1 areas taken by alignments (ft²)	Residential relocation (No.)		
40	4,520,342	485	0	0	4,342	4.62
50	4,328,432	758	0	0	4,302	4.68
60	4,638,662	1,032	0	0	4,340	4.67

Figure 12.10 Sensitivity of optimized alignments to design speed

in the alignment construction, are less sensitive to the geometry of the alignment compared to the other components.

12.4.2   Sensitivity to Design Speed

This analysis tests the sensitivity of solution alignments to the design speed. The design speed is interrelated with many design features of a highway alignment (such as the horizontal curve radius, sight distance, transition curve length, and vertical curve length (crests and sags) of the alignment). In the model, it is specified by model users as an input, and the design features of the solution alignments are computed based on the AASHTO design standards (2001). As shown in Figure 12.10, the model creates smoother and longer horizontal curves at higher design speeds. Of course, the higher design speed also forces the model to generate smooth and long vertical

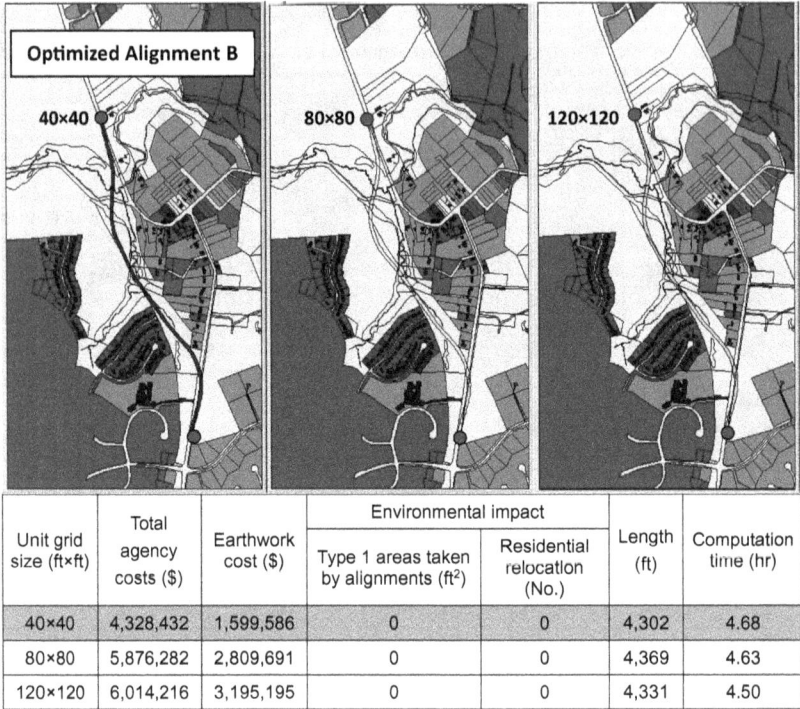

| Unit grid size (ft×ft) | Total agency costs ($) | Earthwork cost ($) | Environmental impact | | Length (ft) | Computation time (hr) |
			Type 1 areas taken by alignments (ft²)	Residential relocation (No.)		
40×40	4,328,432	1,599,586	0	0	4,302	4.68
80×80	5,876,282	2,809,691	0	0	4,369	4.63
120×120	6,014,216	3,195,195	0	0	4,331	4.50

Figure 12.11   Sensitivity of optimized alignments to elevation resolution

curves. This indicates that the model performs correctly in creating highway alignments that satisfy the AASHTO standards.

12.4.3   Sensitivity to Elevation Resolution

Resolution of the input ground elevation may also significantly affect the quality of solution alignments as well as model computation time. This may occur because the rough resolution of the ground elevation may decrease the accuracy of the earthwork cost estimation. As shown in Figure 12.11, there are striking differences in earthwork cost estimation between three optimized alignments generated with different input grid sizes even though they have very similar horizontal profiles; the earthwork cost significantly increases with rough grid size. This indicates that the model may produce unreliable earthwork estimates if the grid sizes are excessive, since terrain

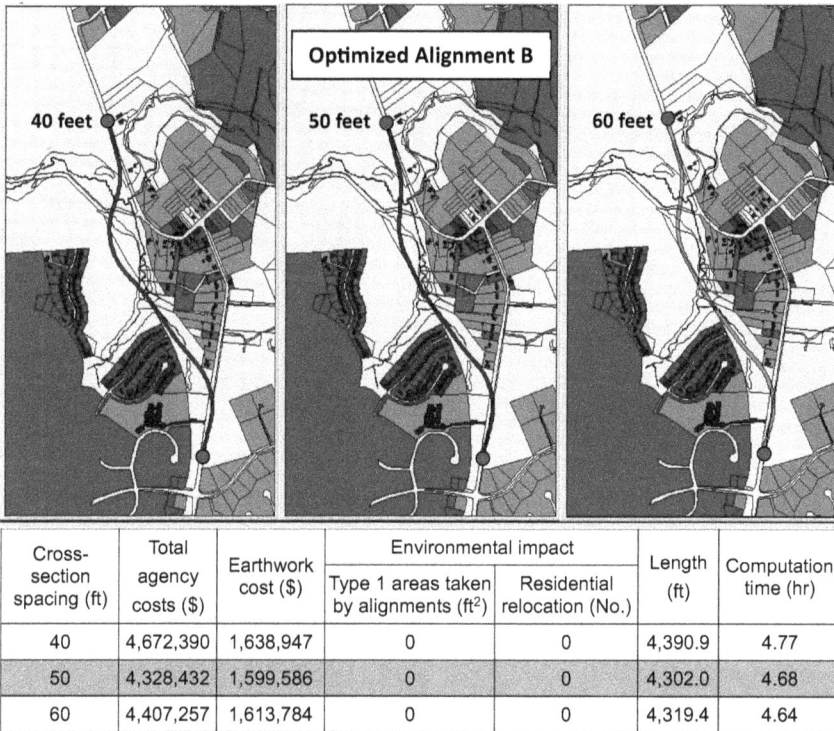

Cross-section spacing (ft)	Total agency costs ($)	Earthwork cost ($)	Environmental impact		Length (ft)	Computation time (hr)
			Type 1 areas taken by alignments (ft²)	Residential relocation (No.)		
40	4,672,390	1,638,947	0	0	4,390.9	4.77
50	4,328,432	1,599,586	0	0	4,302.0	4.68
60	4,407,257	1,613,784	0	0	4,319.4	4.64

Figure 12.12 Sensitivity of optimized alignments to cross-section spacing

elevation estimates may then be too rough. Thus, a fine grid size is recommended in order to estimate the earthwork cost more precisely.

12.4.4 Sensitivity to Cross-section Spacing

Figure 12.12 presents sensitivity to unit cross-section spacing, which is used as the earthwork computation unit of the model. It indicates that the earthwork cost and alignment length can vary depending on the unit cross-section spacing. Note that the cross-section spacing directly influences the precision of earthwork cost computations in the model. Moreover, the alignment length is also affected by the overall earthwork cost since the model seeks to reduce all the considered costs that are affected by the alignment length. In general, however, the variation of earthwork cost due to the differences of cross-section spacing is not significant.

Chapter 13

HAO Model Application in US 220 Project in Maryland

13.1 Project Description

The HAO model has also been applied for the development of alternative alignments of the existing US 220 highway located in western Maryland. The project area of this case study is about 12.5 mile (20 km)-wide and 15.5 mile (25 km)-long, which is significantly larger than the Brookeville Bypass example described in Chapter 12 (see Figure 13.1). The project area is located in the Appalachian Mountains, and thus its ground elevation varies greatly.

Various land uses (such as forest, river, agricultural, and residential areas) exist in the project area, which is mostly covered by forest and cropland. In addition, the project area has many geographically sensitive regions (such as floodplains, state parks, protected lands, and wetlands) that must be carefully considered in selecting the highway location. Many priority funding areas (PFAs), designated by state and federal agencies, are also located in the project area, thus favoring (e.g., with low land cost) candidate alignments which use those areas for their rights-of-way.

The search for the optimized alternative alignment was made within a 4,000 foot (1.2 km)-wide buffer of the existing US 220 between I-68 near Cumberland, Maryland and the West Virginia state line, a distance of approximately 18.6 miles (30 km). The search limit was recommended by a preliminary environmental impact study by MDSHA (2013) which had been prepared as a result of the North South Appalachian Corridor Feasibility Study by the West Virginia Division of Highways (WVDOH, 2006). Note that the WVDOH study showed that the improvement of the existing US 220 has great potential for benefiting the economic development

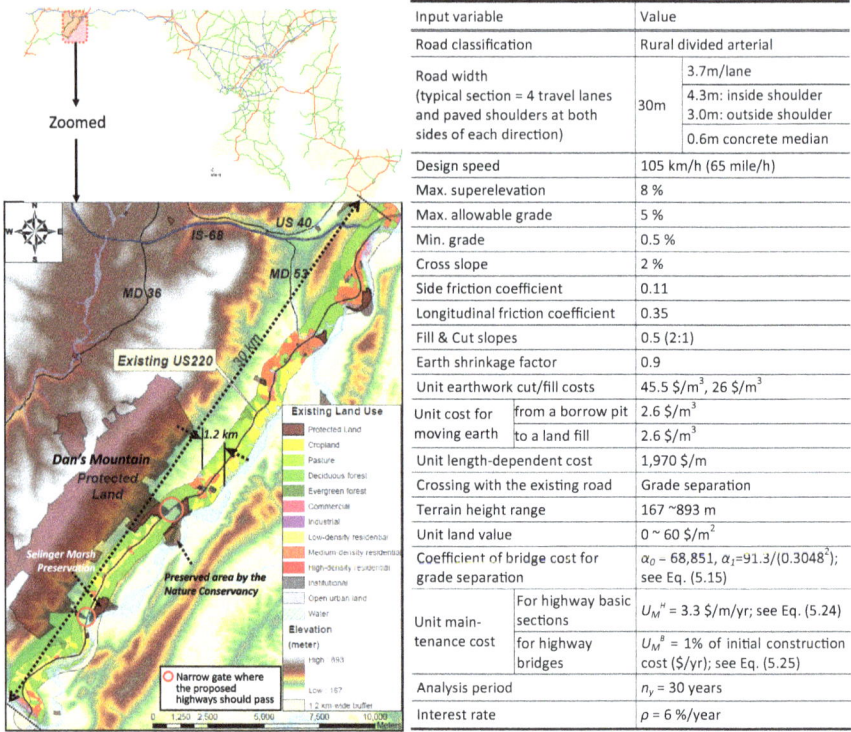

Input variable		Value	
Road classification		Rural divided arterial	
Road width (typical section = 4 travel lanes and paved shoulders at both sides of each direction)		30m	3.7m/lane
			4.3m: inside shoulder
			3.0m: outside shoulder
			0.6m concrete median
Design speed		105 km/h (65 mile/h)	
Max. superelevation		8 %	
Max. allowable grade		5 %	
Min. grade		0.5 %	
Cross slope		2 %	
Side friction coefficient		0.11	
Longitudinal friction coefficient		0.35	
Fill & Cut slopes		0.5 (2:1)	
Earth shrinkage factor		0.9	
Unit earthwork cut/fill costs		45.5 $/m^3, 26 $/m^3	
Unit cost for moving earth	from a borrow pit	2.6 $/m^3	
	to a land fill	2.6 $/m^3	
Unit length-dependent cost		1,970 $/m	
Crossing with the existing road		Grade separation	
Terrain height range		167 ~893 m	
Unit land value		0 ~ 60 $/m^2	
Coefficient of bridge cost for grade separation		$\alpha_0 = 68,851$, $\alpha_1 = 91.3/(0.3048^2)$; see Eq. (5.15)	
Unit maintenance cost	For highway basic sections	U_M^H = 3.3 $/m/yr; see Eq. (5.24)	
	for highway bridges	U_M^B = 1% of initial construction cost ($/yr); see Eq. (5.25)	
Analysis period		n_y = 30 years	
Interest rate		ρ = 6 %/year	

Figure 13.1   Study area and base inputs for the HAO model application to US 220 Project

of Appalachian region. Thus, in this case study we have applied the HAO model to search for the most economical alternative alignments of US 220 within the search limit (i.e., a 0.7 mile (1.2 km)-wide and 18.6 mile (30 km)-long buffer), while preserving environmentally sensitive areas.

The proposed highway is a four-lane rural divided arterial with partial access controls as suggested in MDSHA (2009). Its typical section is 100 foot (30 meter)-wide with a median in the middle and paved shoulders at the both sides of each direction. The design criteria and other important factors employed in this case study are also shown in Figure 13.1. Design of interchange and/or intersection configurations at the northern and southern ends of the proposed highway was not included in the scope of this case study.

13.2 Projection Preparation

13.2.1 Spatial Data

The MDSHA provided various GIS data (such as elevation, land use, and property maps) for use by the HAO model. These GIS data were used to prepare appropriate inputs. Based on those data, four important GIS maps required by the HAO model application were developed: 1) *Land-Use Map*; 2) *Unit Land Cost Map*; 3) *Elevation Map*; 4) *Geographical Constraints Map*.

For a *Land-Use Map*, various spatial data which describe land use, land value, geographical constraints, ground elevation, and location of PFAs were prepared in a GIS data format. Data validation and adjustment were completed at a planning level of detail. To estimate and compare the right-of-way cost (i.e., land acquisition cost) of various alternatives, a *Unit Land Cost Map* was also developed as an input of the HAO model. The *Unit Land Cost Map* was prepared through the integration of various GIS data provided by the MDSHA. As the first step of map integration, digitization of important properties (e.g., property boundary and environmentally sensitive areas) in the project area was processed. Various GIS layers were employed to digitize the important features.

To obtain the unit land value ($/ft^2 or $/m^2) of each property in the project area, a GIS point layer in MdProperty View (which stores property boundary, area, total value, ownership, and address information) was used. The point layer and the digitized property map were superimposed and then merged through the "spatial join" process to calculate the unit land cost of each property in the project area. A 2002 classification of jurisdiction land use/land cover based on a USGS classification scheme was used to classify different land use types in the project area. The land use map of the project area shows more complexity in its northern section due to the dense residential, institutional and commercial areas, while the typical land use in the southern part changes to forests and croplands. Terrain in the northern section is somewhat level or rolling despite the complexity in land use. Toward the south, the terrain becomes more jagged and mountainous. The range of the unit cost is also reasonable due to the elevation and land use condition. Most of the high cost properties are located in the northern section where the relatively smoother terrain and residential land use are accessible.

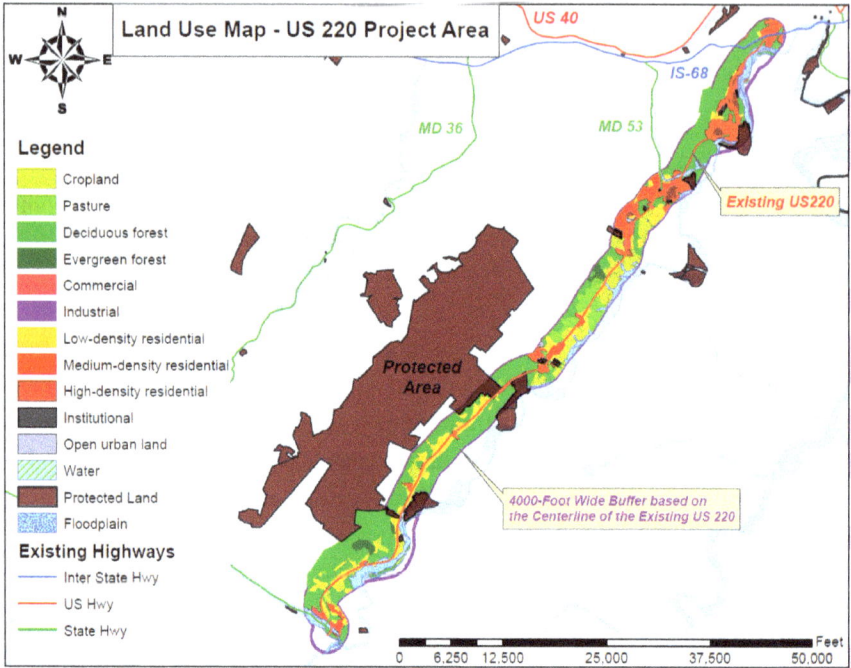

Figure 13.2 Land-use map

1) Land-use Map

Figure 13.2 shows a land-use map developed for this project in which various land use types provided by MdProperty View and three additional types of geographically sensitive regions (i.e., water, floodplains, and protected lands) are included. Due to their relative importance and relevance to the project, some land use types were merged and a total of nine different types were used, as shown in Table 13.1.

2) Unit Land Cost Map

The unit cost of each property is calculated based on the parcel database from the MdProperty View. The total value of all parcel points placed in a polygon was divided by the total area of that polygon. The attribute table of the *Unit Land Cost Map* contains the property ID, unit land cost ($/ft^2 or $/m^2), perimeter, area, and maximum area allowable to be affected by the preferred alignment. Figure 13.3 shows the final *Unit Land Cost Map* developed through the data integration with the GIS data provided by MDSHA.

Table 13.1 HAO model land use classification

Land use	HAO model land use classification
High-density residential	Residential
Medium-density residential	
Low-density residential	
Commercial	Commercial
Industrial	industrial
Institutional	Institutional
Water	Water
Open urban land	Open urban land
Cropland	Cropland
Pasture	Pasture
Evergreen forest	Forest
Deciduous forest	

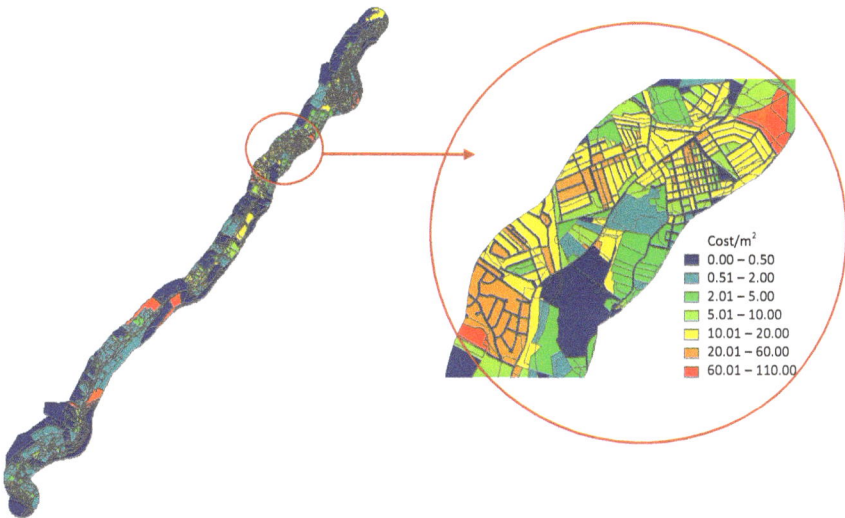

Figure 13.3 Unit land cost map for HAO model application

3) Elevation Map

The elevation map is a raster file containing the ground elevation of the study area. This GIS data was used as an input of the HAO model to estimate the earthwork volume and cost of various possible alignments generated

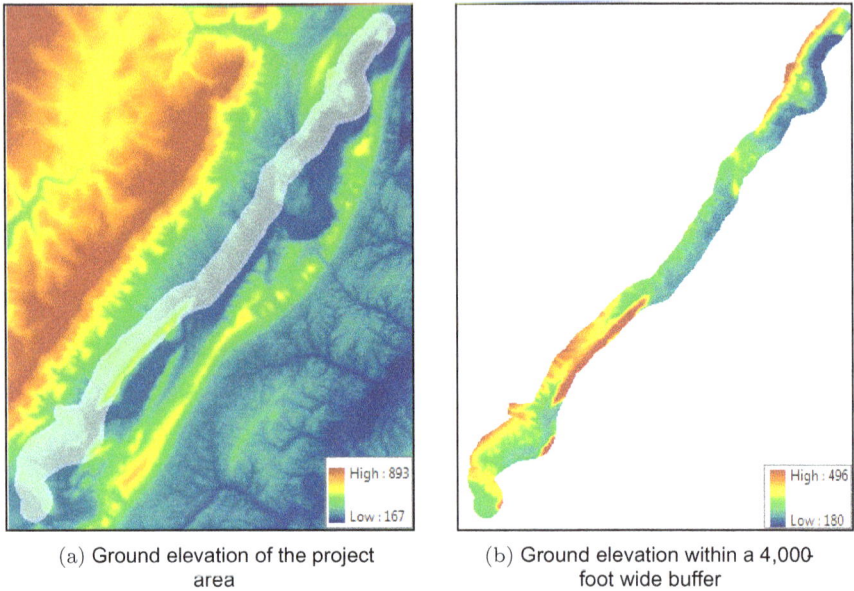

(a) Ground elevation of the project area

(b) Ground elevation within a 4,000-foot wide buffer

Figure 13.4   Elevation map for the HAO model application

from the model. Note that a raster image is a data structure representing a rectangular grid of pixels, viewable via a monitor or other display medium. So, a raster is also called a grid or an image in GIS. The value in each grid cell corresponds to the characteristic of a spatial phenomenon at the cell location. Figure 13.4 shows the ground elevation maps used in the HAO model in this project.

4) Geographical Constraints Map

There are many protected lands in the study area of the US 220 project. These areas are protected for the jurisdiction based on various federal, state, and local government programs. Dan's Mountain State Park (which includes a wide variety of wildlife, mountain streams, and scenic overlooks), Selinger Marsh Preservation area, and environmentally preserved area by the Nature Conservancy are parts of the protected land.

Figure 13.5 shows critical locations in which special care should be exercised in selecting highway alternatives for US 220. As shown in the figure, many protected lands are located in the study area, and they make the alignment search boundary complex. The protected land is regarded as

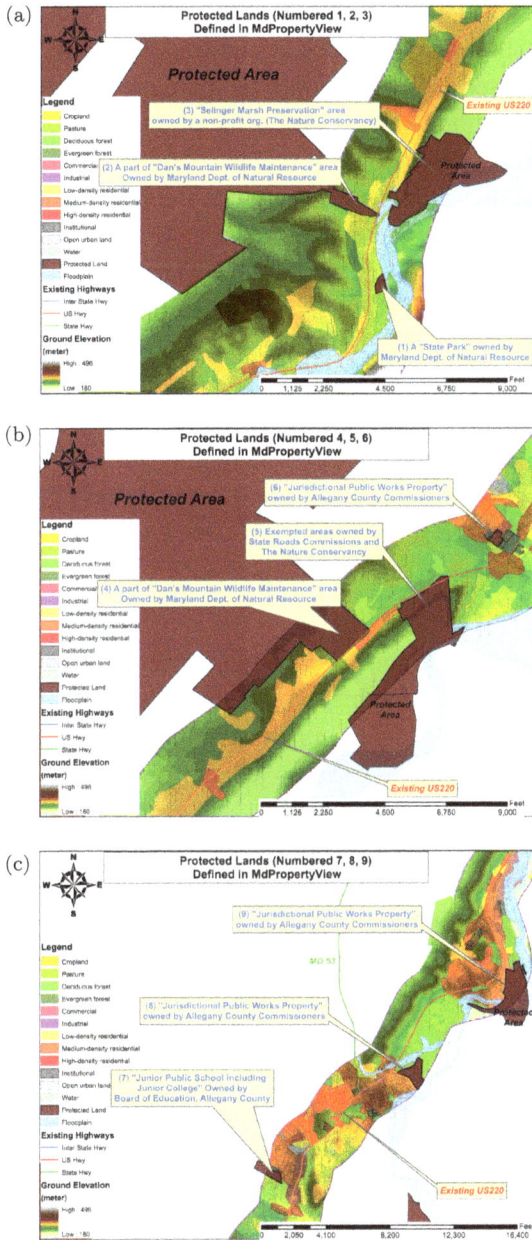

Figure 13.5 Geographical constraints map

an "Untouchable region" in this project so that it should be excluded as a possible location of the proposed highway. Thus, several narrow gates where the proposed highway should pass are pre-specified with reference to the spatial location of the protected land. Such narrow gates will guide the model to avoid generating infeasible alignments that violate the spatial constraints and thus to focus the search on the feasible solutions.

Figure 13.5(a) shows a model search boundary narrowed by State Park, Dan's Mountain Wildlife, and the Selinger Marsh Preservation areas. In this area, widening of the existing US 220 is recommended instead of letting the HAO model search for a new bypass.

Figure 13.5(b) shows another narrow gate of the model search boundary surrounded by Dan's Mountain Wildlife area, exempted areas, and Jurisdictional Public Works Property. Widening of the existing US 220 is also recommended in this area. Figure 13.5(c) shows the location of other protected lands along the study area.

13.2.2   Project Segmentation

Due to the spatial constraints (i.e., protected lands) and complexity of land use in the project area, the highway alternative search space of the HAO model was divided into eight segments (named A, B, C, D, E, F, G, and H). These segments include (1) Case 1 – locations where widening of the existing US 220 is preferred (i.e., segments A, C, E, and H shown in Figure 13.6) and (2) Case 2 – locations where developing new bypasses is recommended (i.e., segments B, D, F, and G shown in Figure 13.7). Note that the start and end points of the new bypass in each section (i.e., diverging and merging points of the new bypass from and to existing US 220, respectively) were obtained, by applying the HAO model to the entire section of the study area. The optimization of the alignments for the new bypass and widening of existing US 220 for each section was then conducted.

Segments A and H are the start and end sections of the project limit where the widening of existing US 220 is recommended due to the complexity in land use and ground elevation. These segments mostly run through pastures, forests, and floodplains; however, commercial and high-density residential areas are also located along existing US 220. Furthermore, ground elevation outside the existing US 220 varies greatly. Thus, a new bypass in these areas may incur significantly high land acquisition costs and earthwork costs.

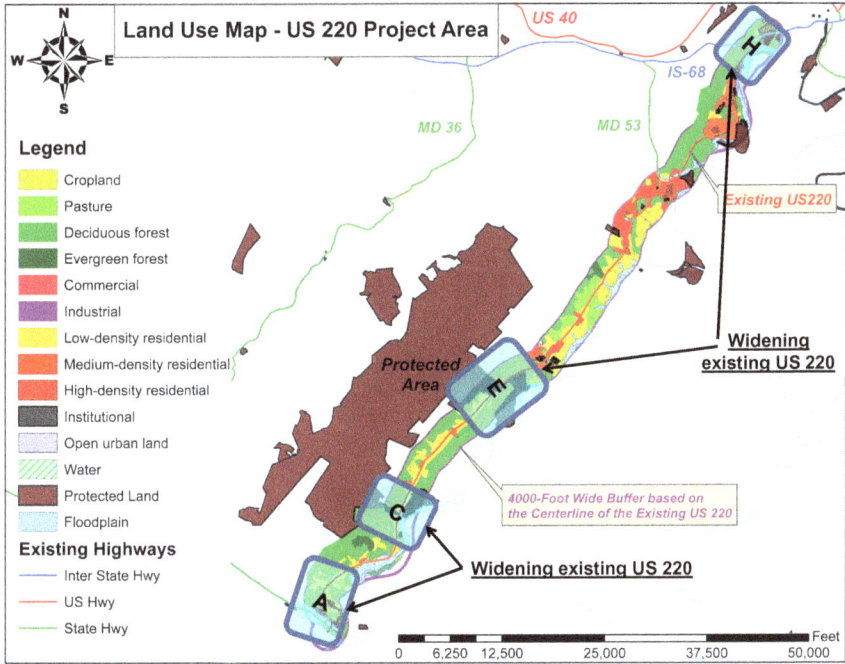

Figure 13.6   A, C, E, and F segments where widening of existing US 220 is preferred

Segments B, D, F, and G are the locations where development of new bypasses is recommended. Sufficient undeveloped spaces are available, and land use is relatively simpler than others. The HAO model was applied to each section to find the best alternative alignment.

Segments C and E are the locations where many protected lands (such as State Park, Dan's Mountain Wildlife Maintenance area, and Selinger Marsh Preservation area) complicate the search for a new bypass. Thus, widening of existing US 220 is strongly recommended to minimize the environmental impact. Road construction costs (such as earthwork costs) may also be significantly reduced by widening the existing road instead of building new bypasses.

13.2.3   Geometric Design Specification

The typical section for the new bypass on segments B, D, F, and G is shown in Figure 13.8. It is called the *Normal Section* (named in the *US 220 Tier*

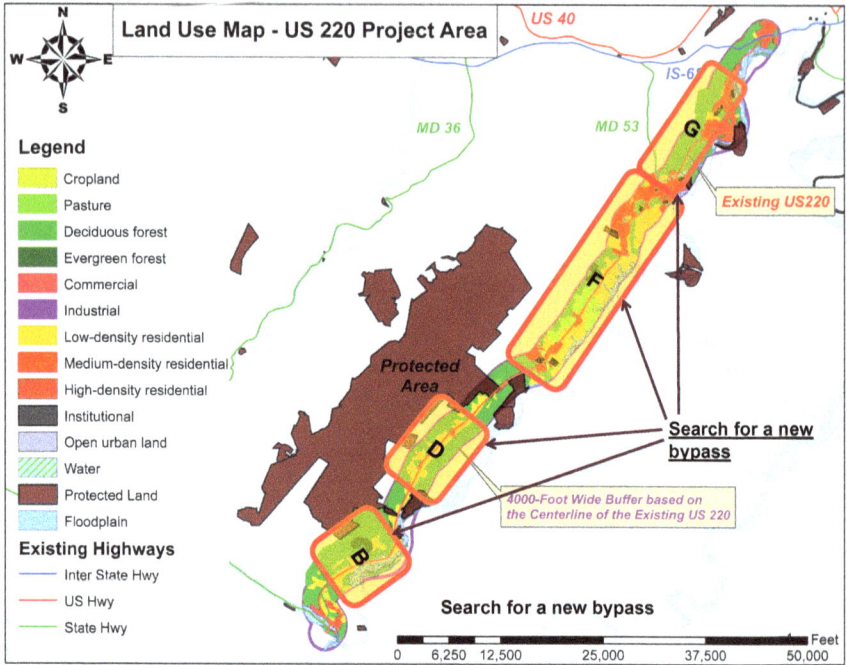

Figure 13.7   B, D, F, and G segments where development of a new bypass is recommended

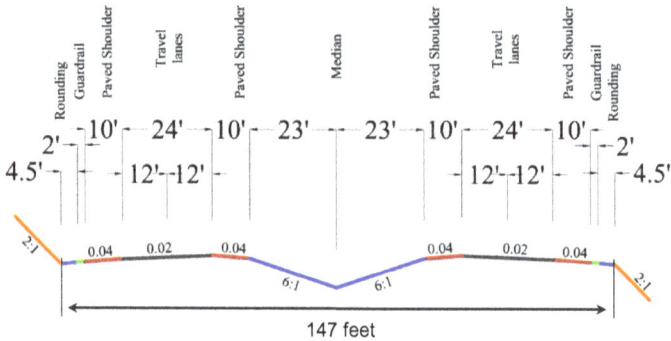

Figure 13.8   Typical section of the new bypass for segments B, D, F, and G

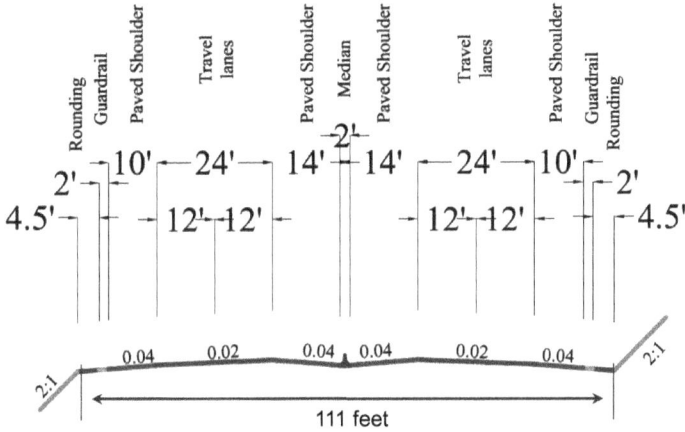

Figure 13.9 Typical section for widening of the existing US 220 in segments A, C, E, and H

One Study by the West Virginia Division of Highways (DOH) and MDSHA, 2006). The total width of the paved section is 98 feet (30 meters), which includes four travel lanes, a concrete median in the middle, and paved shoulders at both sides of each travel direction. Note that the geometric design specification of the new bypass (e.g., design speed, maximum superelevation, and maximum grade) is shown in Figure 13.1, and was used for optimizing alignments of segments B, D, F, and G.

Figure 13.9 shows the typical section for segments A, C, E, and H where widening of existing US 220 is recommended. This typical section is called *Alternate Section* (named in the *US 220 Tier One study*, 2006), and its total width is 134 feet (41 meters) (except guardrail and rounding sections). It includes a wide grass median in the middle and two travel lanes and paved shoulders in each direction.

13.3 Optimization Results

The HAO model was applied to each segment B, D, F, and G to optimize alignments of a new bypass of US 220. For the other segments (i.e., A, C, E, and H), however, the model was only used to estimate the cost of widening the existing US 220. The model searched over 300 generations,

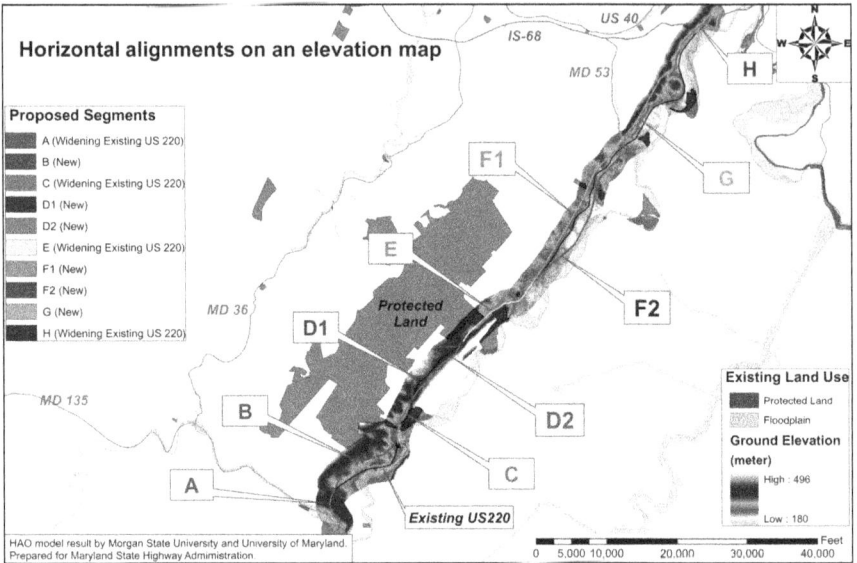

(a) Horizontal alignment on an elevation map

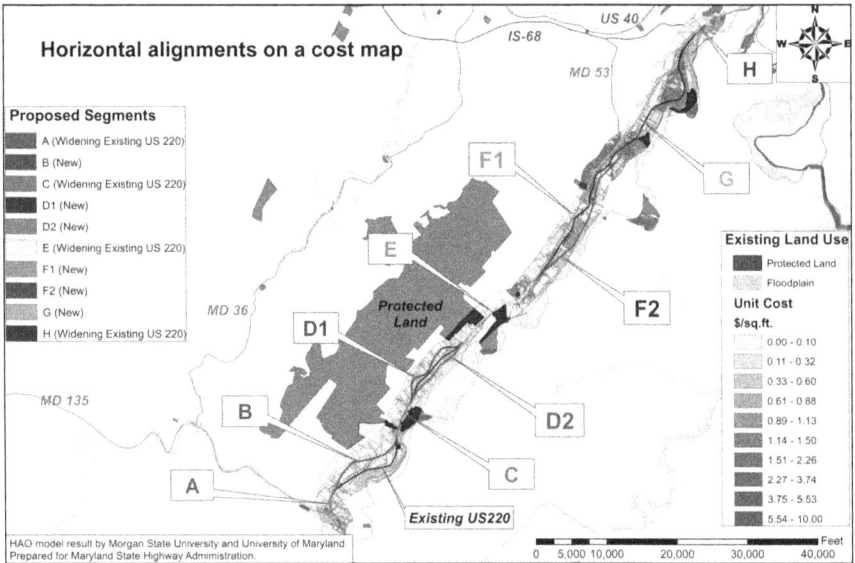

(b) Horizontal alignment on a cost map

Figure 13.10   Optimized alignments of the new bypass for US 220

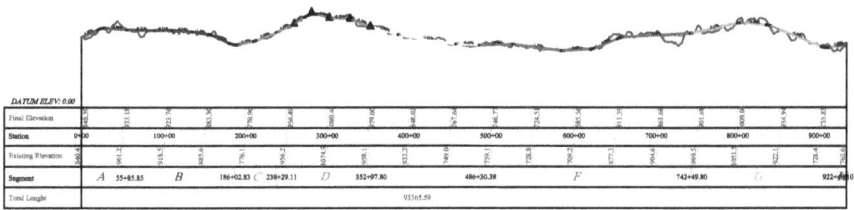

(c) Vertical alignment

Figure 13.10   (*Continued*)

thereby generating and evaluating about 10,000 alternative alignments to find best bypass alignments for each segment B, D, F, and G. Note that the 98 foot-wide *typical section* (shown in Figure 13.8) was used when optimizing alignments of the new bypasses for segments B, D, F, and G, while 134 foot-wide *typical section* (shown in Figure 13.9) was used when widening the existing US 220 for segments A, C, E, and H.

The HAO model objective function used in this project included: length-dependent, right-of-way, earthwork, structure, maintenance, and penalty costs. The assumed design life of the proposed highway was 30 years and a 6 % annual interest rate was used to estimate the present value of the highway maintenance cost. Note that the penalty cost function, Eq. (5.31) described in Chapter 5, was employed to evaluate the amount of environmentally sensitive areas taken by the highway alignments generated from the model. The alignment search was made through 300 generations to obtain the optimized one. It took a relatively longer computation time in this project (about 110 hours with a Desktop PC intel® Core™2 Duo with 2 GB RAM), as compared to the Brookeville Bypass project discussed in Chapter 12, due to the land use complexity and large scale of the project.

The optimized alignment found by the HAO model is shown in Figure 13.10. Its total length is about 18 miles (28.9 km), and consists of eight road segments (A-B-C-D-E-F-G-H) as displayed with different colors. Segments A, C, E, and H share the right-of-way of the existing US 220 but widen it, while segments B, D, F, and G are new bypasses. The horizontal profile of the optimized alignment successfully avoids environmentally sensitive areas (including the protected lands and state parks) and

Table 13.2 Total cost and areas taken by the optimized alignment by land use type

Road segment	Total cost (million $)	Length (m)			Area taken by the proposed road segment (m²) by land use type									Protected land (m²)	Flood-plain (m²)	Existing road (m²)	PFA (m²)
		Total	Basic section	Bridge	Total	Residential	Commercial	Industrial	Institutional	Start Part	Forest	Cropland	Pasture				
A	52	1703	1703	0	139,001	23,972	36,317	0	0	0	44,454	34,258	0	0	14,874	50,781	123,966
B	50	3967	3613	354	268,136	54,975	0	0	0	0	194,230	18,931	0	0	0	18,515	76,933
C	9	1593	1593	0	70,160	65,257	0	0	0	0	4903	0	0	0	8515	41,863	39,084
D	50	3895	3895	0	270,302	150,875	0	0	0	0	89,895	29,532	0	0	0	20,862	185,502
E	54	4064	4064	0	208,491	112,145	3174	0	0	0	92,857	0	314	0	0	63,845	141,853
F	179	7809	6986	823	639,955	230,762	22,107	0	0	0	129,682	204,352	53,053	0	39,265	88,895	621,234
G	222	5486	4785	701	649,170	26,091	59	196	0	0	622,824	0	0	0	30,588	8484	84,573
H	9	341	341	0	32,002	0	0	0	0	0	32,002	0	0	0	32,009	7365	25,598
Total	624	28,857	26,979	1878	2277,216	664,077	61,657	196	0	0	1,210,846	287,073	53,367	0	125,252	300,609	1,298,742

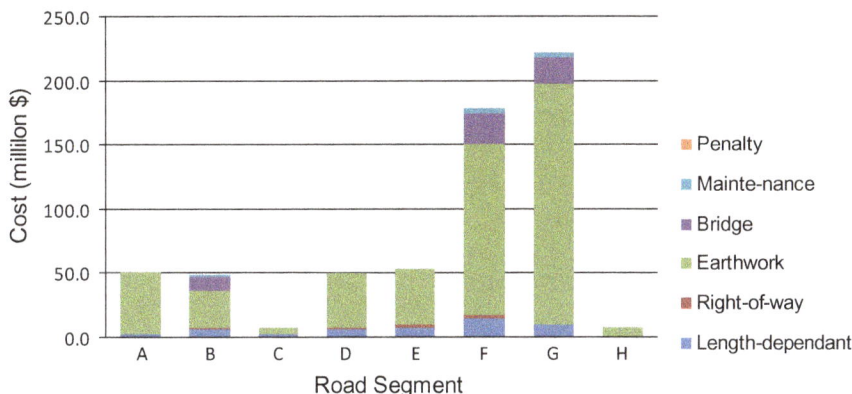

Figure 13.11 Total cost breakdown of the optimized alignment

high land cost areas. However, it unavoidably affects some residential and commercial areas (see Table 13.2) due to the complexity of the land use within the 0.75 mile (1.2 km)-wide buffer. There are four bridges in the entire section of the optimized alignment: one on each of segments B and F, and two on segment G. Among them, the bridge on segment F provides grade separation for the existing US 220, while the others on segments B and G are chosen because bridge construction is more economical there than earthwork. No tunnel is considered in this project.

Figure 13.11 shows the total cost breakdown for the optimized alignment, which is based on the best trade-offs among the six cost components. It shows that the earthwork accounts for a significant fraction (about 80%) of the total cost, which is much more than in the Brookeville case. The bridge structure also accounts for a large fraction of the total cost; however, the fraction of the right-of-way cost is negligible, because the US 220 project area is mountainous and property values in it are relatively low. Other costs such as user cost, contingency cost, and utility relocation cost are not considered in this case study, and thus the total cost may be underestimated. It is noted that among the eight road segments, segments B, F, and G account for more than 72% of the total cost, and cover about 60% of the entire section. Earthwork, bridge, and length-dependent costs are the three major agency costs for these segments; however, the maintenance cost also constitutes a substantial fraction.

Chapter 14

HAO Model Application to Maryland ICC Project

14.1 Project Description

14.1.1 Overview of the ICC Study

Maryland Route 200, also known as the Intercounty Connector (ICC), was proposed as a multi-modal transportation improvement to help address traffic needs between the I-270/I-370 and I-95/US-1 corridors within central and eastern Montgomery County and northwestern Prince George's County in the State of Maryland (see study area map in Figure 14.1). Many local, state, and federal agencies as well as consultant companies have been working cooperatively to facilitate the progress and effectiveness of the ICC project. According to the draft environmental impact statement and major investment study of ICC project (FHWA and MDSHA, 1997), the need for the ICC is based on the following factors:

- *"The I-270 corridor, which is one of the premier highway facilities providing direct cross-count routes in the State of Maryland, has only one access-controlled highway linking it to the I-95 Corridor. I-95 not only has extensive existing and planned development straddling it throughout the corridor between Washington and Baltimore, but also serves to connect the Washington Metropolitan Area to Baltimore and the entire northeast United States.*
- *The one access-controlled link connecting I-270 and I-95 is I-495 (the Capital Beltway), which is currently operating at capacity during peak periods, causing many persons traveling between the I-270 and I-95 corridors to utilize the local roadway system instead. These roads are not designed or intended to carry this longer distance travel. Furthermore, the Beltway is at the southern perimeter of the ICC study area and therefore does not provide a direct cross-county route for traffic in this area.*

211

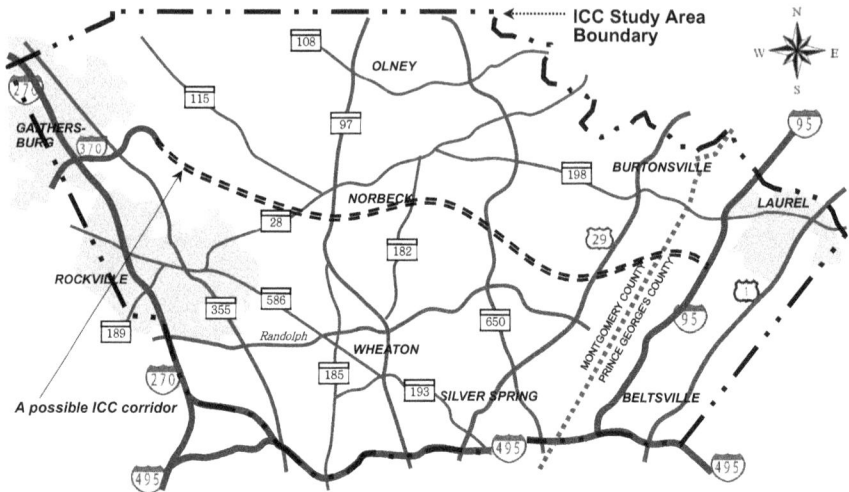

Figure 14.1   ICC study area

Source: FHWA-MDOT (1997).

- *Numerous roadways within the study area currently operate at or near capacity and have fairly high accident rates due to the many entrances and intersections.*
- *There is a lack of continuous east-west express transit service.*
- *The number of trips within the ICC study area, especially east-west trips, is expected to increase substantially in coming years.*
- *The number of intersections and roadway links in the ICC study area operating at or near capacity is also expected to increase substantially."* (FHWA and MDSHA, 1997)

Given such needs, the purposes of the ICC are to:

- *"Connect the existing and planned development areas between and adjacent to the two corridors with I-270 and I-95.*
- *Connect, in an environmentally responsible manner, the I-270 and I-95 corridors and accommodate, safely and efficiently, the east-west transportation movements between the corridors.*
- *Relieve congestion on existing roads not meant to accommodate cross-county traffic".* (FHWA and MDSHA, 1997)

14.1.2 Description of HAO Model Application to ICC Project

The ICC is a large-scale transportation improvement project in terms of time, space, and funding. Various critical factors (such as political, environmental, geographical, and even capital investment issues) are interrelated, and vast amounts of data and resources are required for the problem. The highway alignment optimization model is also applied to this project in order to identify the best alternatives for the ICC; furthermore, the model's network level optimization capability is demonstrated through this case study. A problem description and the assumptions defined for this case study are presented in the following.

(A) Problem Description

In this case study, not only the highway alignments themselves but also their two endpoints and cross-points with existing roads are simultaneously optimized throughout the model application. Furthermore, traffic improvements due to the addition of the new alignments on the existing road network are also considered in the optimization process besides the other major alignment sensitive costs. Thus, the model objective function employed for the ICC application is the sum of (i) total user cost saving and (ii) total agency cost; i.e., $\Delta C_{User} + C_{T\_Agency} = (\Delta C_T + \Delta C_V) + (C_L + C_S + C_M + C_R + C_E)$. Note that the accident cost (C_A), which is another component of the user cost, is suppressed from the model objective function in this application.

(B) Assumptions and Limitations

Only major highways (at least State level) are selected for specifying the existing road network, which is required for the traffic assignment process. Two continuous search ranges for the start and end points of new alignments are assumed to be known (along the I-370 and I-95, respectively), and a trumpet-type interchange is considered at each endpoint. Three major highways (MD-97, MD-650, and US-29) run between the two endpoints, and they are almost unavoidably intersected by the new alignments. For linking those major highways and new alignments, four different types of crossing-structures are considered (listed below), and the best structure type for each cross-point is determined during the optimization process.

- 4leg at-grade intersection
- Grade separation
- Clover interchange
- Diamond interchange

Traffic operating on the ICC study area in a base year (i.e., a base year O/D trip matrices) is known and increases annually with a given growth rate. A roughly digitized horizontal map is used here because preparation of a detailed GIS map is relatively quite expensive for model application to the large-scale project.

14.2   Input Data Preparation

14.2.1   Road Network

Twenty major State highways are selected to represent the existing road network of the ICC study area (see Figure 14.1). These highways are used to construct a network incidence matrix, which is used for an input of the traffic assignment process. Note that the incident matrix is kept updated during the optimization process if newly generated highways are added to the existing road network. Characteristics of the major highways, such as number of lanes, capacity, and speed limit are presented in Table 14.1.

14.2.2   Traffic Information

(A) Zonal Descriptions
According to the Metropolitan Washington Council of Government (MWCOG), the Washington Metropolitan area is divided into 2,191 Traffic Analysis Zones (TAZ) consisting of parts of Maryland, Washington DC, and Northern Virginia. Among them, 423 TAZs, which are possibly affected by the new ICC construction, are selected for the model application as shown in Figure 14.2. Note that 198 TAZs identified as the immediate ICC impact areas by the MDSHA (Clifton and Mahmassani, 2004) are included in the selected TAZs (see Table 14.2).

(B) O/D Trip Matrices
The year 2010 is assumed to be the base year for the model application to ICC study. Two types of modes (auto and truck), and three time periods (AM-peak, PM-peak, and Off-peak) are considered. The base year O/D trip

Table 14.1 Characteristics of major highways in the ICC study area

Road name	No. of lanes	Speed limit	Capacity per lane	Access control*
I-95	8	65	2200	Full
I-270/I-495	12	60	2200	Full
I-495/I-370	8	60	2200	Full
US-29	6	55	2000	Partial
MD-198	4	55	2000	General
MD-183/MD-185/MD-193/MD-197/MD-198	6	45	1800	General
MD-201/MD-355/MD-586/MD-650/MD-97				
MD-28/US-1	4	45	1800	General
MD-182/MD-189	4	40	1800	General
MD-198/MD-97	2	45	1800	General
MD-108	2	40	1800	General

*Notes:

- Full: fully access controlled highways without use of at-grade intersections; only interchanges and grade separations are used.

- Partial: partially access control highways with mixed use of grade separations, interchanges, and at-grade intersections.

- General: no access controlled highways.

matrices for different modes and different time periods were obtained from the MWCOG. Note that O/D tables for different trip purposes (e.g., home-based work, home-based shopping, and non-home-based work trips) are not considered in this case study. As shown in Figure 14.2, 33 trip production/attraction points (i.e., centroids) are heuristically identified (mostly) at the ends of the existing highways. These points are designed to aggregate O/D trip pairs between the selected TAZs. Each point represents several TAZs near it (i.e., its corresponding TAZs are identified based on the distance from it). Thus, 178,929 (423 × 423) O/D pairs are aggregated to 1,089 (33 × 33) pairs. The O/D trip matrices used in this case study are summarized in Appendix B.

14.2.3 GIS Map Preparation

An alignment search space is specified within the ICC study area, as shown in Figure 14.3. The total area size of the search space is about 108,362.4

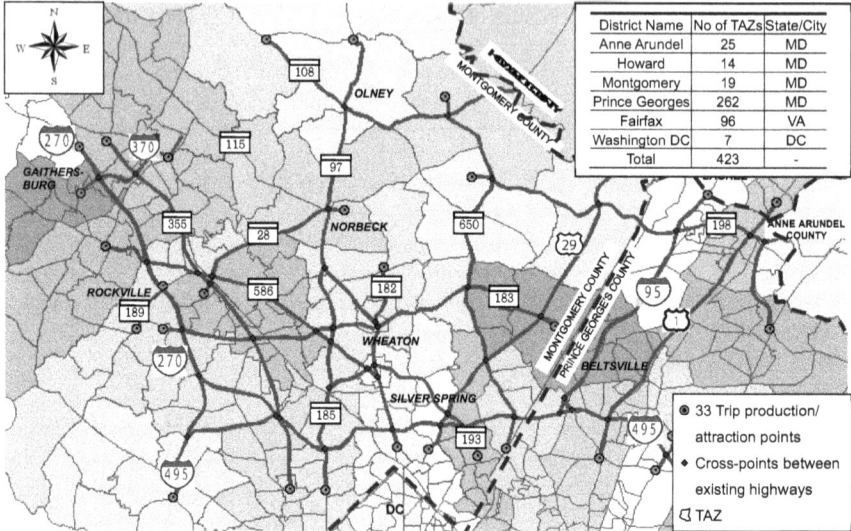

Figure 14.2   Selected TAZs for model application to ICC project

Table 14.2   Immediate ICC impact area by TAZ and jurisdiction boundary

County name	Zone number	No. of TAZs	State
Montgomery	394~468, 473~509, 526~556, 577~582, 585~592	157	MD
Prince Georges	781~792, 865~891	39	MD
Howard	1083	1	MD
Anne Arundel	1091	1	MD
Total	—	198	—

acre (16.6 mile long and 10.2 mile wide). The search spaces for the start and end points of the new alignments are identified along the I-370 and I-95, respectively, as shown in the figure. The Euclidean distance between the start and end points is approximately 14.44 mile (23.243 km).

(A) Horizontal Map

Through a horizontal map digitization process, more than 2,000 geographic entities (including rivers, parks, wetlands, existing highways, and residential and commercial properties) are represented as polygons; these retain their unique property information (such as, spatial location, area, and property value).

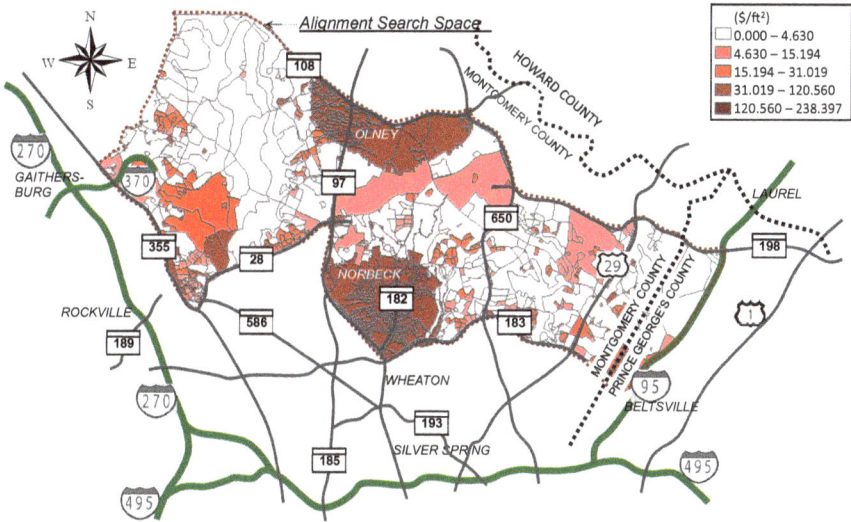

(a) Unit Property Costs in the Selected Search Space

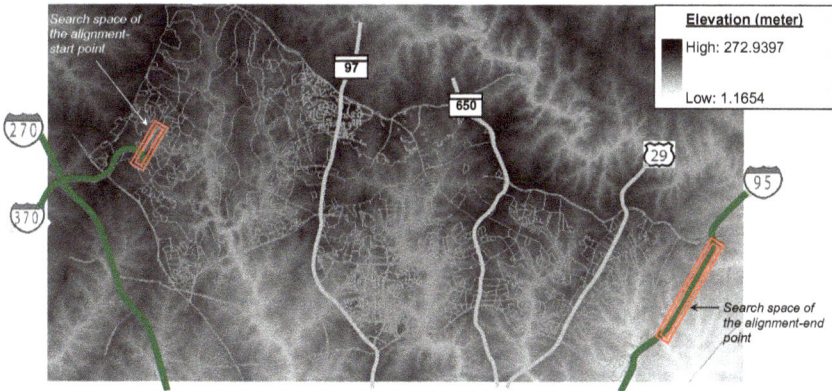

(b) Ground Elevation in the Selected Search Space

Figure 14.3 Alignment search space selected for ICC case study

The unit cost ($/ft^2 or $/m^2) of each property in the search space is obtained from MDProperty View 2003. Note that the horizontal map of the search space (shown in Figure 14.3(a)) is somewhat more roughly digitized here rather than for the Brookeville case study because digitizing all detailed geographic entities (e.g., all individual building structures) in the large-scale project area is very expensive. Thus, environmental impact summaries and right-of-way costs of the alignments resulting from the model application to the ICC study may not be as accurate as those for the Brookeville case study.

(B) Ground Elevation Map

A Digital Elevation Model (DEM), which provides ground elevations of the ICC study area with a grid base, was downloaded from United States Geological Survey (USGS) website as shown in Figure 14.3(b). Note that the DEM is used to calculate the alignment earthwork cost in the model. In the DEM, the study area ground elevations are divided into evenly spaced grids of size 30 meters × 30 meters (103 feet × 103 feet). Finer grids may be selected for precise earthwork calculation as desired. The elevation range in the ICC study area is 4 to 895 feet (1 to 273 meters). The darker areas represent higher elevations.

14.2.4   Important Input Parameters

Input variables required for computing the alignment-sensitive costs in the model's application to the ICC study are summarized in Table 14.3. These include road width and design speeds as (i) agency cost variables, and annual traffic growth rate and truck percentage in the traffic as (ii) user cost variables. Regarding the agency cost variables, the design speed of the proposed alignment of the ICC is initially set to 60 mph (96 kph), and its cross-section is assumed to represent an 8-lane major highway with a 106 foot width (11 feet lanes and 9 feet shoulders). Cross-section spacing for the earthwork calculation is set to 50 feet (15 meters) and the minimum vertical clearance for crossing with existing highways is assumed to be 15 feet (4.5 meters).

Regarding the user cost variables, the annual traffic growth rate and truck percentage in the traffic of the ICC study area are assumed to be 10% and 15%, respectively. In addition, the interest rate and analysis period are set to 3% and 5 years, respectively. Unit travel time value ($/hr), average vehicle occupancy (person/veh), and fuel prices ($/gallon) for autos and trucks are presented in Tables 10.3, 10.4, and 10.6 of Chapter 10, respectively. Please refer to Table 14.3 for the other important input parameters used in the ICC case study. Note that values of all these input variables should be cautiously defined because they may sensitively affect the resulting alignments.

Table 14.3 Baseline inputs used in the model application to the ICC case study

	Input variables	Value
Agency cost variables	No. of intersection points (PI's)	8~12
	Road width	106 foot, 8-lane road (11" lane, 9" shoulder)
	Design speed	60 miles/h (96 km/h)
	Maximum superelevation	0.06
	Maximum allowable grade	5 %
	Coefficient of side friction	0.16
	Longitudinal friction coefficient	0.28
	Distance between station points	50 feet (15 m)
	Fill slope	0.4
	Cut slope	0.5
	Earth shrinkage factor	0.9
	Unit cut cost	35 $/yard^3 (45.5 $/m^3)
	Unit fill cost	20 $/yard^3 (26 $/m^3)
	Cost of moving earth from a borrow pit	2 $/yard^3 (2.6 $/m^3)
	Cost of moving earth to a fill	3 $/yard^3 (3.9 $/m^3)
	Unit length-dependent cost	400 $/feet (656 $/m)
	Terrain height ranges	4~895 feet (1~273 m)
	Unit land value in the study area	0~238 $/ft^2 (0~2,562 $/m^2)
	Structure type on the start and end points	Trumpet interchanges
	Structure types on the cross-points with existing highways	Grade separation, 4-leg at-grade intersection, Clover and Diamond interchanges
User cost variables	Traffic growth rate	5%
	Truck percentage in the traffic	5%
	Interest rate	3 %
	Analysis period	5 years
	Base year O/D	2010 O/D trip matrices (see Appendix B)
	Unit travel time value	9.28 $/hr for auto drivers (50% of average wage rate);16.84 $/hr for truck drivers (see Table 10.3)

(Continued)

Table 14.3 (*Continued*)

Input variables	Value
Average vehicle occupancy	1.550 persons/auto; 1.144 persons/truck (see **Table 10.4**)
Fuel prices	2.73 $/gallon for auto; 2.67 $/gallon for truck (see **Table 10.6**)
Number of major highways used for constructing an existing road network	20 (at least State level)
Number of centroids (trip production/ attraction points) pairs	1,089 (=33 × 33)

14.3 Optimization Results

14.3.1 Determination of Traffic Reassignments

Figure 14.4 shows example solution alignments possibly generated at the beginning of the model search process (called initial population stage). Such initial population members may include straight alignments as well as some possible candidate alignments selected based on judgments of highway designers and planners. The solutions are improved over successive generations with the aid of the customized genetic operators (Jong, 1998) and the efficient solution search methods (described in Chapters 7 and 8 for FG and P&R approaches, respectively) after the initial population stage is completed.

Recall that the bi-level optimization feature of the HAO model (see Chapter 10) is designed (i) to update the configuration of the road network after a new highway alignment is generated, and next (ii) to find equilibrium traffic flows in the updated network from the traffic assignment process, and (iii) finally to evaluate the total cost of the new highway (including the user cost savings as well as agency cost) associated with its construction. It should be noted, however, that the bi-level optimization may not be efficient in cases when the assignment results for the networks updated with different highway alternatives are very similar. For instance, the difference in traffic volumes which would operate on the new highways, shown in Figure14.4,

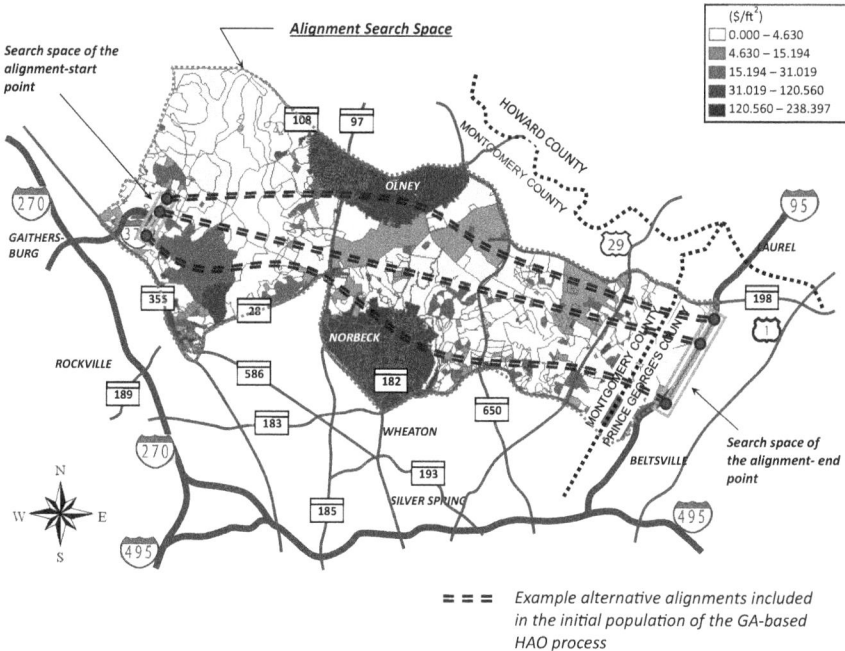

Figure 14.4 Alignment search space with example alternatives included in the initial population of HAO process for the ICC case study

may be negligible although their start and end points as well as horizontal (and even vertical) alignments generated from the model significantly differ. In such a case, processing the traffic assignment (i.e., finding equilibrium traffic flows) for every updated network with the new alternative generated is wasteful. The *Preprocessed Traffic Assignments* is developed here for determining whether the bi-level optimization feature is needed during the optimization procedure for given problems. Note that this procedure is pre-processed with sample solutions generated at early stages of the alignment optimization including initial population. Please refer to Figure 14.5 for the *Preprocessed Traffic Assignments* procedure. The notations used in the process are presented in Table 14.4.

Preprocessed Traffic Assignments Procedure

(1) STEP 1: Generate initial population, including straight and curved alignments.

Domain of the
highway start

B₁

B₂

9 segment pairs

A₁ – B₁ A₂ – B₁ A₃ – B₁

A₁ – B₂ A₂ – B₂ A₃ – B₂

B₃

A₁

A₂

A₃

Domain of the
highway endpoints

Existing

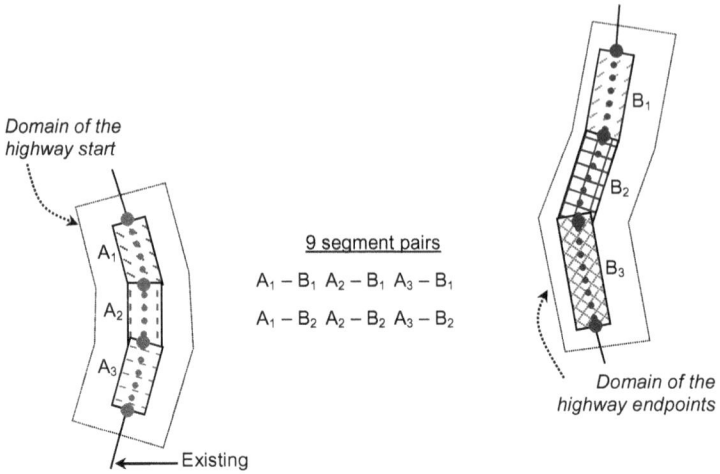

(a) Disaggregate domains for highway endpoints generation

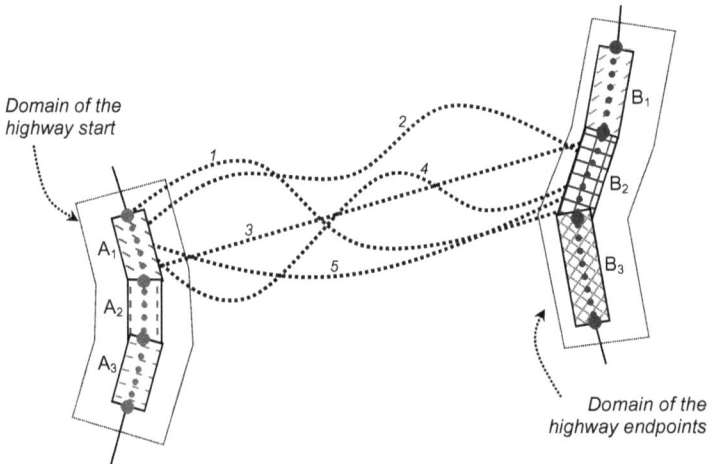

Domain of the
highway start

2

1

B₁

4

B₂

3

5

A₁

B₃

A₂

A₃

Domain of the
highway endpoints

(b) Example highways generated with road segment pair A₁–B₂

Figure 14.5 Road segment pairs specified for preprocessed traffic assignment

- STEP 1-1: Identify domains of highway endpoints specified for the endpoint generations.
  - As shown in Figure 14.5(a), domains of the highway start and end points are divided into three road segments each in the ICC case study (i.e., $n_{seg1} = 3$, $n_{seg2} = 3$).

Table 14.4 Notation used in the *Preprocessed Traffic Assignments* procedure

Notation	Descriptions
$F_{TA} =$	user-specifiable threshold value for determining whether the bi-level optimization feature is needed
$N_{ipop} =$	total number of sample highways generated in the initial population; $N_{ipop} \geq 5 \times n_{seg1} \times n_{seg2}$
$n_{seg1}, n_{seg2} =$	total number of the road segments specified for start and end points of a new highway alignment, respectively
$x_{new\_CV} =$	coefficient of variation of the predicted traffic volumes operating on the initial population members; $x_{new\_CV} = x_{new\_SD}/x_{new\_M}$
$x_{new}j =$	predicted traffic volumes that would operate on the jth new highway of the initial population
$x_{new\_M}, x_{new\_SD} =$	mean and standard deviation of all $x_{new}j$, respectively for $j = 1, \cdots, N_{ipop}$

- STEP 1-2: With each pair of road segments, generate sample alignments including straight and curved alignments.
  - In the ICC case study, 9 (= 3 × 3) segment pairs are identified for the endpoints generation, and at least five sample alignments are generated with each segment pair. Thus, more than 45 (= 5 × 9) highway alternatives are generated during the initial population stage (see Figure 14.5(b)).

STEP 2: Find $x_{new}j$ for all j ($j =$1, ..., N_{ipop}) from the traffic assignment process, where $x_{new}j$ = predicted traffic volumes that would operate on the jth new highway of the initial population, and can be found through the traffic assignment process for the network updated with the new highway addition. Note: N_{ipop} = total number of sample highways generated in the initial population ($N_{ipop} \geq 5 \times n_{seg1} \times n_{seg2}$).

STEP 3: Compute x_{new_M} and x_{new_SD}, where x_{new_M} and $x_{new\_SD} =$ mean and standard deviation of all $x_{new}j$ ($j = 1, ..., N_{ipop}$), respectively.

STEP 4: Compute x_{new_CV}, where x_{new_CV} = coefficient of variation[1] of the predicted traffic volumes ($x_{new}j$) operating on the initial population members; $x_{new\_CV} = x_{new\_SD}/x_{new\_M}$

---

[1] A small coefficient of variation indicates that the assignment results for different alternatives are relatively consistent.

STEP 5: Check whether $x_{new\_CV} \leq F_{TA}$ or $x_{new\_CV} > F_{TA}$, where F_{TA} = a user-specifiable threshold value for determining whether the bi-level optimization feature is needed. (Note that we assume $F_{TA} = 0.05$ in the ICC case study.)

- If $xx_{new\_CV} \leq F_{TA}$ (i.e., the traffic assignment results are relatively consistent for the initial population):
  → Stop the traffic reassignment procedure for alternatives generated over the successive generations. Instead, the results of the preprocessed traffic assignments with the initial population will be used for estimating the user costs of those solutions.

- Otherwise (if $x_{new\_CV} > F_{TA}$):
  → Keep processing the traffic reassignments (beyond the initial population stage) with additional alternatives generated until F_{TA}^{th} generation (see STEP 6). Note that K_{TA} is a user-specifiable parameter, which is set to 50 generations in the ICC case study.

STEP 6: During the F_{TA} generations, compute $x_{new\_CV}^{i}$ for alternatives generated with the specified road segment pairs ($l = 1, \ldots, n_{seg1} \times n_{seg2}$).

- Recall that $9\ (= n_{seg1} \times n_{seg2})$ pairs of road segments are specified in the ICC case study (see Figure 14.5(a)), and the traffic reassignment results are saved for each segment pair during the K_{TA} generations:
  → Compute $x_{new\_CV}^{1}, \ldots, x_{new\_CV}^{9}$ and
  → Check whether $x_{new\_CV}^{i} \leq F_{TA}$ or $x_{new\_CV}^{i} > F_{TA}$ for all i

If $x_{new\_CV}^{i} \leq F_{TA}$, the traffic reassignment results will be used for estimating the user cost of other alternatives generated with the corresponding road segment pair during the remaining generations. Otherwise ($x_{new\_CV}^{i} > F_{TA}$), the traffic reassignment will be processed for all alternatives generated over the successive generations.

Figure 14.6 shows the predicted traffic volumes (veh/hr) which would operate on the sample ICC alternatives generated at the initial population stage. About 50 alternatives are generated in the initial population of the ICC case study. This result indicates that the traffic volumes operating on the various alternatives are relatively similar (with $x_{new\_CV} = 0.0415$) despite

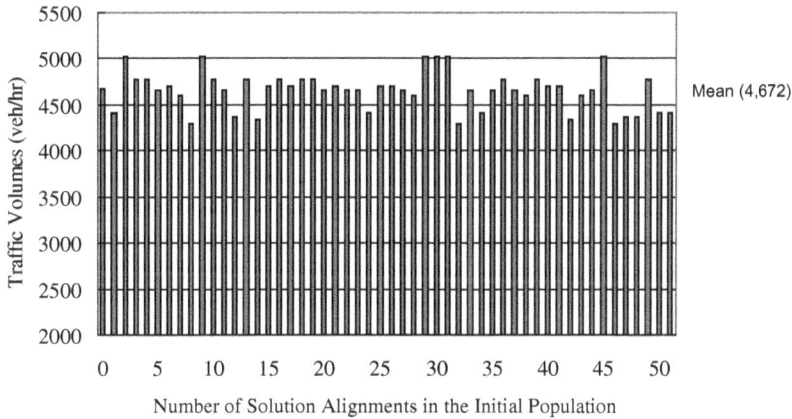

# of lanes*	Mean ($x_{new\ M}$)**	Standard deviation ($x_{new\ SD}$)	Coefficient of variation ($x_{new\ CV}$)
8	4,672	194.0258	0.0415

*The proposed alignment is assumed to be a 106 foot wide, 8-lane road (refer to **Table 14.3**)
** Average hourly traffic (average of AM, PM, Off-peak volumes) operating on the new alignments

Figure 14.6 Predicted traffic volumes operating on the new alignments of the initial population for the ICC case study

their different start and end points locations and different horizontal profiles. Thus, the traffic reassignments are not performed through the successive model search processes (after the initial population stage); instead, the preprocessed assignment results are used for estimating the user costs of the alternatives generated during the rest of the search process.

14.3.2 Optimized Alignments

The alignment optimization model searches over 300 generations (including the initial population stage) to find the cost effective ICC alternatives given the input data shown in Table 14.3. A desktop PC, Pentium Dual CPU (3.0 GHZ, 3.0 GHZ) with 2 GB RAM is employed to run the model, and about 8,400 alignments are evaluated during the search process. It takes a relatively long time (about 24 hours) to run through 300 generations because the ICC study area is very large (16.6 mile long and 10.2 mile wide) and contains many geographic entities.

(b) Horizontal alignment

(b) Vertical profile

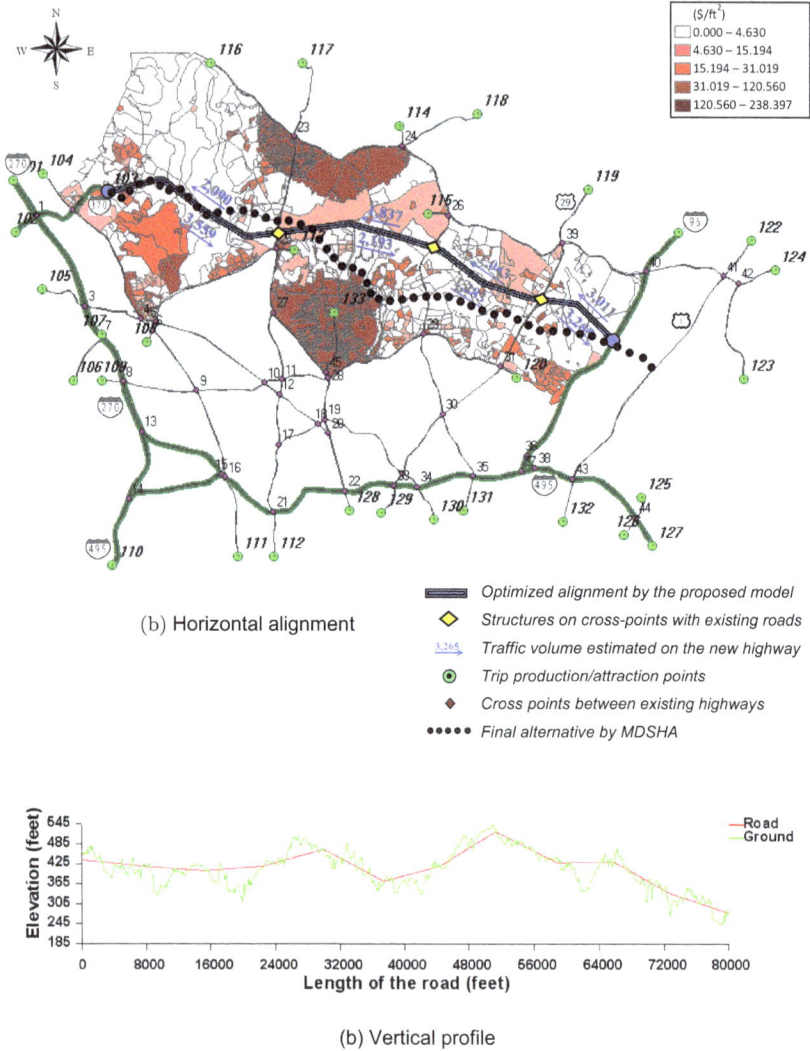

Figure 14.7 Horizontal and vertical profiles of optimized alignment for the ICC case study

Note that the model runs five times (searching over 300 generations in each) with different input PI's (8 to 12 PI's). As a result, the optimized solution found with 10 PI's seems the most preferable due to its lowest objective function value, although alignment profiles and objective function values of all the five solutions are very similar. Figure 14.7 shows horizontal and vertical profiles of the optimized alignment obtained after 300 generations with 10 PI's (i.e., the most preferable one). As shown in the figure, the optimized alignment has seven horizontal curves that satisfy the given design standards, while avoiding the predefined control areas and high cost properties. In addition, its vertical alignment closely follows the ground elevation, while minimizing its earthwork cost. The new highway is 16.02 miles long. Three clover interchanges and two trumpet interchanges are built for facilitating turning movements at the crossing points with existing roads.

It is important to note here that the resulting alignment is found based on the model inputs provided in Table 14.3; thus, some limitations in data accuracy may apply here. The resulting alignment may be further improved or changed if more precise and detailed inputs (such as more detailed environmental consideration and O/D traffic information) are provided. For the readers' information, the ICC alternative finally proposed by the MDSH is also shown in Figure 14.7. The figure also presents average traffic volumes (veh/hr) which would operate on the new alignment in the base year (2010). These results are calculated from the traffic assignment process of the highway network updated with the new highway construction. Expected traffic improvements on three existing major highways (I-95, I-495, and I-495) due to the new highway construction are summarized in Table 14.5. The results indicate that traffic conditions on I-95 can be significantly improved after the system development (with 26% traffic reduction). Traffic on I-495 and I-270 can also be improved (with 18% and 8% reduction, on average) with the aid of the new highway. Note that the input O/D trip matrices used for the assignment are presented in Appendix B.

(A) Change in Objective Function Value Over Successive Generations

In order to assess the behavior of the objective function over the successive generations, the objective function values are plotted at various generations, as shown in Figure 14.8. It is observed that the objective function values in the first few generations are extremely high. However, the value drops considerably until about 79 generations. The improvement in the objective function value becomes very slow (almost negligible) after that. The final objective function value is about 16 million, which is reached at the 160th generation.

Table 14.5   Average traffic on major interstate highways before and after the new alignment construction (2010 base year)

Major highways		Traffic volume (veh/hr)		
		No build	Build	Reduction (%)
Optimized	West Bound	—	2,575	—
Alignment	East Bound	—	1,995	—
I-95	South Bound	5,858	4,349	26
	North Bound	6,125	4,549	26
I-495	West Bound	6,050	4,908	19
	East Bound	5,749	4,767	17
I-270	South Bound	7,966	7,406	7
	North Bound	7,349	6,721	9

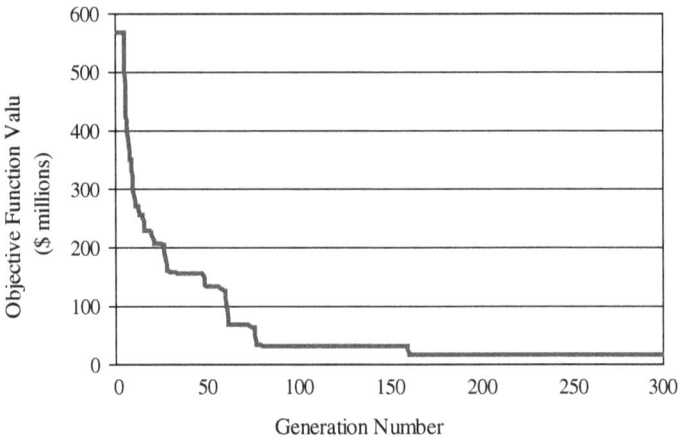

Figure 14.8   Changes in objective function value over successive generations for ICC study

(B) Fraction of Various Costs

It is noted that since the objective function consists of the seven cost components (travel time, vehicle operation, earthwork, right-of-way, length-dependent, structure, and maintenance costs), the proposed optimization model attempts to find the best trade-off between the various cost components and obtain the minimum total while satisfying the specified geographical and design constraints. An analysis is performed to investigate the percentage of various costs in the model objective function value for the

Table 14.6 Cost breakdown of the optimized alignment for the ICC case study

Type of cost		Millions ($)	Percentage (%)
Total agency costs	Earthwork	266.29	59.84
	Length-dependent	33.83	7.60
	Right-of-way	10.92	2.45
	Structures	131.08	29.46
	Maintenance	2.87	0.65
Subtotal		449.99	100.00
Total user cost savings	Travel time	−296.78	69.18
	Vehicle operation	−132.19	30.82
Subtotal		−428.98	100.00
Total costs (Objective function value)		16.08	

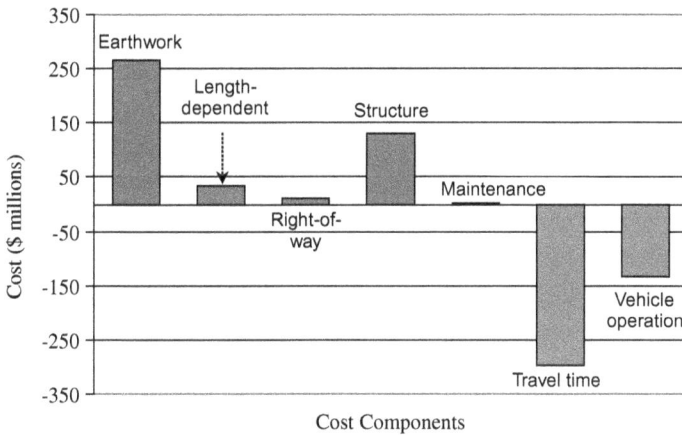

Figure 14.9 Comparison of various costs for optimized alignment of ICC case study

optimized alignment found for the ICC case study. A detailed breakdown of its objective function value is shown in Table 14.6, and Figure 14.9 shows the fractions of the various costs.

The results indicate that (i) travel time cost saving, which can be obtained from the system improvement and (ii) the earthwork cost required for the new road construction, make up the first and second highest fractions of the total objective function value, respectively. They dominate the other cost components included in the objective function. The vehicle operating cost

Table 14.7   Sensitivity of objective function value to analysis period

Unit: ($) millions as of 2010 base year			
Analysis period	Total cost	Total agency cost	Total user cost saving
1	360.27	442.67	−82.40
2	276.81	443.27	−166.46
3	191.64	443.86	−252.22
4	104.72	444.43	−339.72
5	16.01	444.99	−428.98
6	−74.51	445.53	−520.04
7	−166.90	446.05	−612.95
8	−261.17	446.56	−707.73
9	−357.37	447.05	−804.42
10	−455.55	447.53	−903.07

saving and the structure cost also account for large fractions of the total objective function value. These results suggest that care should be taken in using appropriate cost functions in the optimization model to reflect all important costs, although most highway agencies in the field tend to ignore the user costs in the road planning phases. Note that the negative values of the user cost savings indicate that the user costs estimated before the system improvement are greater than those after the road construction. The impacts of the optimized alignment on environmentally sensitive regions of the ICC study area are not presented here since the input land-use maps (which provide spatial locations of various environmentally important features of the study area) used in the ICC case study are not detailed and precise enough.

(C) Sensitivity to Analysis Period

A sensitivity analysis is performed to observe when the user cost savings due to the new highway development exceed the total agency cost required for its initial construction and periodical maintenance. The variation of the total cost (i.e., objective function value) with respect to different analysis periods are presented in Table 14.7. The result indicates that five years after the new road construction the user cost savings exceed the total agency cost so that the total cost becomes negative value, which means that the highway development project starts to benefit from the year 2016.

(a) Distribution of objective function values

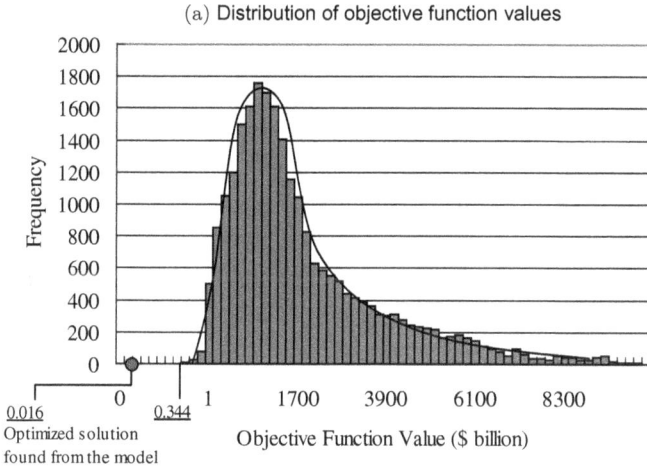

0.016
Optimized solution
found from the model

Objective Function Value ($ billion)

(b) Descriptive statistics of objective function values (unit: $ billion)

	Min	Max	Mean	Median	Standard Deviation
Random Search	0.344	9955	2065	1390	2113
Model	0.016				

Figure 14.10 Comparison of solutions found from random search and optimization model for the ICC case study

14.3.3 Goodness Test

Recall that the ICC case study is quite a different model application compared to the Brookeville example. Highway endpoints as well as its alignments are simultaneously optimized, and traffic improvements on the existing road network due to a new highway development are considered together with various highway agency costs in the ICC application. Thus, a statistical analysis is also performed here to test the goodness of the best solution found by the model.

A set of sample solutions (30,000) is randomly generated to compare them with the optimized solution found by the model. It is observed that the best solution of the random sample yields an objective function value 344 million, while the objective function value of the worst one is 9,955 billion. The sample mean is about 2,065 billion and the standard deviation is 2,113 billion. A distribution diagram of the random sample and its descriptive statistics are presented in Figure 14.10. The relative position of the optimized solution found is also indicated on that figure. The results show that the sample distribution has an offset of 344 million, which is much higher than the optimized solution (16 million) found by the model. This means that the optimized solution dominates all the sample solutions; it is 21 times smaller than the best of 30,000 randomly generated solutions. Such results give us confidence that the optimized solutions found by the model are really excellent when compared to other possible solutions to the problem.

Chapter 15

Related Developments and Extensions

This chapter indicates that many additional developments, enhancements, and new applications remain to be pursued by interested researchers. Thus, various exciting developments in highway location and alignment optimization and related research fields may be expected in the coming years.

15.1 Related Developments

The HAO models presented in the previous sections have been specifically developed to support the planning and alignment design of new roads and relatively small road networks. These models are already quite powerful, comprehensive, and flexible in their capabilities. For example, they have been used to widen and improve existing roads while limiting deviations from the previous road alignments, as shown in Chapter 12 regarding the Brookeville Bypass Road and in Chapter 13 for a study on US 220 in Maryland.

The experience gained in developing the HAO models inspired the authors and their collaborators to apply some HAO concepts and methods to other transportation modes. Thus, fairly similar methods have been used to develop alignment evaluation and optimization methods for rail transit lines (Jha *et al.*, 2007; Lai and Schonfeld, 2012 and 2016), intercity passenger railroads (Kang *et al.*, 2014) and mountain railroads (Li *et al.*, 2016 and 2017; Pu *et al.*, 2019a and 2019b). In Lai and Schonfeld (2012) and (2016), three-dimensional alignments of rail transit lines are jointly optimized with station locations along them while vertical alignments between rail transit stations are optimized to account for train energy use, among other factors. Energy considerations favor "dipped" alignments which raise stations so that gravity assists trains in accelerating downhill away from stations and decelerating uphill toward stations. Such designs had been analyzed earlier in Kim and Schonfeld, (1997), but without GIS data and only for vertical alignments.

The work on mountain railway alignments was mainly contributed by a research team from Central South University (CSU) in Changsha, China, led by Professor Hao Pu. That team has explored alternative solution methods including distance transforms (two and three dimensional), particle swarm algorithms, and hybrid methods that include genetic algorithms. The CSU team has also researched the related problem of re-creating railroad alignments from imperfect ("noisy") field measurements (Li *et al.*, 2019). This becomes important if the track alignment is distorted over time by the passage of vehicles and the original design is not available anymore. Even if the original design is available, it may actually be more economical to revise the distorted alignment slightly to satisfy safety constraints rather than implement major changes required to duplicate the original alignment. It should be noted that the problem of "re-creating" or "identifying" existing alignment is also the subject of research for highways (Bassani *et al.*, 2016, Higuera de Frutos and Castro, 2017).

GIS-based evaluation and optimization methods for locating other kinds of transportation facilities or other kinds of infrastructure are still surprisingly rare. In an M.S. thesis of 2014, Tong Zhao made an interesting effort to optimize the location of airport facilities (Zhou 2014; Zhou and Schonfeld, 2015). It seems that the methods that have been already developed for optimizing alignments for roads, railways, and rail transit lines could readily be adapted to the optimization of transportation facilities such as pipelines, canals, aerial cable cars, hyperloops, and magnetic levitation routes, as well as to route-based facilities such as electric power lines. They may also be applied to temporary routes such as aircraft flight paths that consider air traffic restrictions, noise, winds and weather conditions, or ship routes that consider currents, waves, and storms.

The potential for applying GIS-based optimization methods for facility location seems great and untapped. This field of research has been dominated by mathematical programming and heuristic methods that treat facilities as shapeless and dimensionless points to be located at nodes of graphs. Current computation capabilities and GIS database qualities should enable more ambitious and realistic location methods. With appropriate evaluation methods, GIS data could support location of multiple similar (e.g., fire-houses, stores, restaurants) or hierarchically-related facilities (e.g., production, storage, and retail outlets in logistic networks).

The HAO algorithms presented in this monograph, which was originally developed for optimizing fixed alignments for highways, have also been adapted for optimizing routes of robotic agents on search and supply missions through hazardous three-dimensional terrain (Jha *et al.*, 2008; Chen *et al.*, 2010; Kang *et al.*, 2011). These or other algorithms might also be applied in optimizing the paths of ground vehicles through difficult terrain or flight paths that must consider winds, noise effects on the ground and air traffic control restrictions.

15.2 Extensions

While the HAO models developed to date are already practically applicable and have extensive capabilities, many further improvements and extensions of capabilities may be considered. In this section some of the potential improvements are briefly discussed. Some of the improvements discussed in the following have already been incorporated in some form within existing versions of HAO models but considerable room for improvement remains. The list of potential improvements includes the following:

1. Interactive design optimization
   - This improvement would involve human decision makers in the search loop
2. Applications of other AI and hybrid algorithms
   - The potential of distance transforms (DT's) and particle swarm optimization (PSO), possibly combined with genetic algorithms, seems considerable
3. Distributed processing
   - Genetic algorithms and population-based algorithms in general are naturally well-suited for parallel processing of different members of one generation on different processors
4. Improved specification of objectives, preferences, and weighting factors associated with political requirements and preferences

5. Improved specification of geographic constraints, e.g. on environmentally or historically sensitive areas to be avoided or locations to be closely approached

6. Detailed specification of trial alignments

   - This improvement may help guide the search near pre-selected alignments and speed up convergence to optimized solutions. However, convergence speed has not been a significant problem to date

7. GIS integration improvements

   - These improvements may include integrating GIS from multiple sources, dealing with different GIS resolutions, and managing defects and gaps in GIS data

8. User interface improvements

   - Such improvements would make it easier for beginners to import input data and employ the models while allowing more experienced users to efficiently conduct different kinds of analyses (such as analysis of sensitivity to inputs, objectives and constraints, evaluation of user-specified alignments, or optimization within very restricted rights-of-way)
   - Users would also effectively visualize resulting alignments as well as present detailed output characteristics and performance measures with these improvements
   - These improvements would also include embedded statistical tests of solution quality, as initiated in Jong and Schonfeld (2003)

9. Automated preparation of inputs

   - This improvement may considerably reduce the time and effort needed to compile the geographic inputs into GIS data bases

10. Bounded corridor search enhancements

    - This improvement includes extreme cases where a road may be widened within narrow bounds, while keeping nearly the same centerline
    - Feasible gates (FGs) tighter than those introduced in Chapter 7 of this monograph may be considered to optimize adjustments and widening of existing roads

11. Optimization of highway alignments with varying design specifications, such as longer than minimum curves, spirals and three-center curves

12. Multi-objective Pareto optimization; such optimization techniques have already been initiated in a paper by Yang *et al.* (2014)

13. Sequential hierarchical analysis: zooming on decreasing sections; such analysis was initiated in Kim's Ph.D. dissertation at University of Maryland (Kim, 2001) and a paper by Kim *et al.* (2005)

14. Sequential analysis of constraints and deficiencies in a solution search process

    • Computation efficiency can be significantly improved by identifying constraint violations and possibly remedying them before candidate solutions are subjected to more comprehensive evaluation. Such improvement was initiated in a study by Kang *et al.* (2009)

15. Improved guidelines to determine critical parameters of the model, such as:

    • Guidelines for variable PI density based on terrain and land-use complexity
    • Guidelines for search parameters such as population size, operator mix, and stopping criteria
    • Guidelines for constraint and weight parameters

16. Improved interfaces with existing roads, rail lines, rivers, and mountains

    • This would extend the studies of Kim (2001) and Kim *et al.* (2007) where intersections, interchanges, bridges, and tunnels are incorporated in the highway alignment optimization process

17. Integrating traffic simulation into the models:

    • A simulation-based approach (e.g., VISSIM or PARAMIX) may be linked to the models for more precise user cost estimation. However, it should be noted that because of a heavy computational burden expected, the simulation-based approach may not be necessary for all generated alternatives but only for several candidate alternatives (e.g., a set of the best ones found in every generation)

18. Integration of alignment design with construction planning and scheduling

19. Extensions of the highway alignment optimization problem to larger networks, beyond those analyzed in Kang (2008)
20. Improved tradeoffs between initial and operating costs

    - This would especially consider how gradients and curves affect operating speeds, energy use and accident costs. Such an improvement was initiated in a study by Kang *et al.* (2013)

21. Improved sight distance analysis in the alignment generation and evaluation process

    - This may include the incorporation of the sight distance analysis, as formulated in Lovell (1999) and Lovell *et al.* (2001), into the model

22. Improved safety models with human factors analysis
23. Improved analysis of tradeoffs between initial and recurring maintenance costs; such analysis was initiated in Jha and Schonfeld (2003)
24. Analysis of pollutant emissions dispersal and resulting effects on affected population and environment; such analysis was initiated in a paper by Mishra *et al.* (2014)
25. Analysis of road space allocation and managed lanes in highway alignment optimization
26. Analysis of surface geologic hazards, such as rock falls and mud slides in highway alignment optimization
27. Analysis of terrain effects on shadows, and hence on iced or wet road surfaces in highway alignment optimization
28. Accessibility analysis, i.e. the travel time and cost of accessing candidate alignments
29. Noise analysis and mitigation in highway alignment optimization; such improvement has already been initiated by Jha and Kang (2009)
30. Adaptations for driverless vehicles (100% driverless or mixed vehicle fleets) in highway alignment optimization
31. Optimizing highway alignments for long tunnels based on imperfect geologic data
32. Scanning alignments to assess and remedy weaknesses, especially regarding safety and maintenance costs
33. Consideration of new highway effects on future traffic growth and mix as well as future land use and development

34. Optimizing ranges of uncertainties regarding unit costs, traffic volumes, and other factors
35. Coordination of vertical and horizontal alignments, according to AASHTO design guidelines (AASHTO, 2011)
36. Analysis of indirect benefits (e.g., incomes, employment, development, competitiveness, etc.) in the process of evaluating alternative alignments
37. Inclusion of other relevant analyses in highway alignment optimization, such as:

- Variable cross-sections, including widening at curves
- Variable speed limits
- Hydrologic analysis
- Wetlands compensation
- Separate carriageways
- Elevated and depressed sections
- Retaining walls
- Disruptions during construction
- Accessibility during construction
- Local optimization of curves
- Marking and signing
- Guardrail placement
- Bicycle lanes, sidewalks, and driveways
- Turning lanes
- Passing lanes
- Climbing lanes
- Fill material choices
- Lighting
- Sun glare
- Adaptations for other transportation modes.

Appendix A

Notation Used in the Monograph

Table A.1 Notation

Notation	Description
α_0, α_1	Coefficients used in bridge cost computation
$\beta_{E^0}, \beta_{E^1}, \beta_{E^2}$	Coefficients used in computing C_{PE}
$\beta_{HG_0}, \beta_{HG_1}, \beta_{HG_2}$	Penalty parameters for user-defined geographical constraints
$\beta_{HR^0}, \beta_{HR^1}, \beta_{HR^2}$	Coefficients used in computing C_{P_DHR}
$\beta_{HS^0}, \beta_{HS^1}, \beta_{HS^2}$	Coefficients used in computing C_{P_DHS}
$\beta_{ST^0}, \beta_{ST^1}, \beta_{ST^2}$	Coefficients used in computing C_{P_DST}
$\beta_{VG^0}, \beta_{VG^1}, \beta_{VG^2}$	Coefficients used in computing C_{P_DVG}
$\beta_{VL^0}, \beta_{VL^1}, \beta_{VL^2}$	Coefficients used in computing C_{P_DVL}
$\beta_{VS^0}, \beta_{VS^1}, \beta_{VS^2}$	Coefficients used in computing C_{P_DVS}
$\gamma_0, \gamma_1, \gamma_2$	Coefficients used in additional tunnel cost computation
$\omega_0, \omega_1, \omega_2$	Binary integers used in earthwork cost computation
ρ	Annual interest rate
θ_{CP}	Intersection angle between two cross roads
θ_{EP}	Intersection angle at the endpoint of the new highway
θ_{max}	Maximum allowable deflection angle
θ_{PI_i}	Deflection angle at the ith point of intersection (PI)
θ_{ST_i}	Spiral angle at the ith horizontal curved section
θ_{vc}	An angle between a vertical cutting line and the X axis of a given map
δ_i	The center point of the ith horizontal curved section
$\delta_{a,k}^{rs}$	Indicator variable (1: if arc a is on path k between O/D pair r-s; 0: otherwise)
$\| \ \|$	Norm (or length) of a vector
$\cdot$	Inner (dot) product used in vector operation
A	Arc set of a given highway network; $\mathbf{A}^0, \mathbf{A}^1$ = a set of arcs in the road network before and after a new highway is added, respectively; $a^0 \in \mathbf{A}^0, a^1 \in \mathbf{A}^1$

(Continued)

241

Table A.1 (*Continued*)

Notation	Description
A_k^j	Area of property piece k affected by alignment j
A_c, A_f, A_{tc}, A_{tf}	Cross-sectional areas under cut, fill, transitional cut, and transitional fill conditions
A_k	Affected area of the kth land parcel by the highway alignment generated
A_{pIS}	Intersection pavement area
A_T	Total area of a land parcel affected by the highway alignment
c_a'	Practical capacity of arc a (defined as 80% of actual capacity)
C_A	Present value of total accident cost
C_{a^T}	Additional tunnel cost which includes cost for ventilation and lighting
$C_{B^{IC}}$	Small bridge cost for grade separation of the existing road at the interchange
C_E	Earthwork cost for a new highway alignment development
$C_{E^{IC}}$	Interchange earthwork cost
$C_{E^{IS}}$	Intersection earthwork cost
C_{E^T}	Tunnel earthwork cost
C_{EN}	Environmental cost for a new highway alignment development
C_H	Total haul cost for earthwork cost estimation
C_L	Length-dependent cost for a new highway alignment development
C_M	Present value of total maintenance cost for a new highway alignment development
C_{M^H}	Present value of the maintenance cost for highway basic segments
C_{M^B}	Present value of bridge maintenance cost
$C_{p^{IC}}$	Interchange pavement cost
$C_{p^{IS}}$	Intersection pavement cost
C_P	Total penalty cost of a new highway alignment
C_{PE}	Penalty associated with environmentally sensitive areas taken by the new highway
C_{PD^H}	Penalty cost for violating design constraints of horizontal alignments
C_{PD^V}	Penalty cost for violating design constraints of vertical alignments
$C_{PD^{HR}}$	Penalty cost for violating the minimum horizontal curve radius
$C_{PD^{HS}}$	Penalty cost for violating the minimum horizontal sight distance
$C_{PD^{ST}}$	Penalty cost for violating the minimum length of spiral transition curve
$C_{PD^{VL}}$	Penalty cost for violating the minimum length of vertical curve
$C_{PD^{VS}}$	Penalty cost for violating the minimum vertical sight distance

Table A.1 (*Continued*)

Notation	Description
C_{P_DVG}	Penalty cost for violating the maximum allowable gradient
C_R	Right-of-way cost for a new highway alignment development
$C_{R^{IC}}$	Interchange right-of-way cost
$C_{R^{IS}}$	Intersection right-of-way cost
C_S	Total structure cost for a new highway alignment development
C_{S^B}	Small bridge cost for grade separation
$C_{S^{IC}}$	Interchange cost
$C_{S^{IS}}$	Intersection cost
C_{S^T}	Tunnel cost
C_{Total}	Total cost for a new highway alignment development
C_T	Present value of total travel time cost
ΔC_T	Present value of total travel time cost saving after a new highway development in the existing road network for the analysis period
C_{T_Agency}	Total agency cost for a new highway development
C_{T_User}	Total user cost for a new highway development
$C_{T\_User}^0, C_{T\_User}^1$	Total user costs before and after a new highway construction, respectively
$C_{T\_B}^0, C_{T\_B}^1$	Total travel time cost in the base year (\$/yr) over the road network before and after a new highway is added, respectively
ΔC_V	Present value of total vehicle operating cost saving after a new highway development in the existing road network for the analysis period
C_V	Present value of total vehicle operation cost
$C_{V\_B}^0, C_{V\_B}^1$	Total vehicle operating cost in the base year (\$/yr) overt the road network before and after a new highway is added, respectively
$\mathbf{CS}_i$	The point of change from circle to spiral pertaining to $\mathbf{PI}_i$
D_{offset}	Additional offset for horizontal feasible gate
Df_i^h	Horizontal tangent deficiency between $\mathbf{PI_i}$ and $\mathbf{PI_{i+1}}$
Df_i^h	Vertical curve-length deficiency at between VPI_i and VPI_{i+1}
D_{VPI_i}	Distance between VPI_i and VPI_{i+1}
d_i	The coordinate of the intersection point at the ith vertical cutting line
d_{iU}, d_{iU}	The upper and lower bounds of d_i, respectively
E, E'	Environmentally sensitive and insensitive areas, respectively
E_k	A dummy variable to indicate whether property piece k is inside E; ($E_k = 1$ if k is inside E; $E_k = 0$, otherwise)
End_V	Endpoint of a vertical alignment; $End_V = (H_{n+1}, Z_{n+1})$, where H_{n+1}, is alignment length and Z_{n+1} is given

Table A.1 (*Continued*)

Notation	Description
EP	Start or end points of a highway alignment; $\mathbf{EP}_1 = \mathbf{PI}_0$; $\mathbf{EP}_2 = \mathbf{PI}_{n_{PI}+1}$
E_{VIS}	Earthwork volume required for the at-grade intersection
F_A	Average accidents (frequency) estimated
F_i^q	q^{th} horizontal feasible gate for generating i^{th} PI; F_i^q can be defined with S_i^l, S_i^{l+1}, and D_{offset}; $q = 1, \ldots, m_i/2$
F_{TA}	User-specifiable threshold value for determining whether the bi-level optimization feature is needed
f_j	Fuel consumption rate of jth mode operated on the new highway
f_k^{rs}	Flow on path k connecting O/D pair $r-s$, $\mathbf{f}^{\mathbf{rs}} = (\cdots, f_k^{rs} \cdots)$
$\mathbf{f}_{a^0}^{AM}, \mathbf{f}_{a^0}^{PM}, \mathbf{f}_{a^0}^{OFF}$	Vector representations of unit vehicle operating cost for traffic (autos and trucks) on arc a^0 during AM, PM, and Off peak hours, respectively before a new highway is added in the road network
f_{a_Auto}, f_{a_Truck}	Fuel consumption of auto and truck, and can be estimated with their average speeds on arc a; $, a \in \mathbf{A}$
g_i	Forward or back tangent grade at the ith vertical curve section
g_{max}	Maximum allowable gradient
H	A vector of duration of different time frames per year; $\mathbf{H} = [H_{TAM}, H_{TPM}, H_{TOFF}]$; $H_{TAM}, H_{TPM}, H_{TOFF}$ = total number of AM, PM, and OFF peak hours in a year, respectively
H_i	H coordinate of VPI_{ji}, for $i = 1, \ldots, n_{PI}$
HFB, HFB'	Horizontal feasible and untouchable areas, respectively
h_m	Minimum vertical clearance
$\mathbf{I}_{E^i}, \mathbf{I}_{E^{i+1}}$	A pair of intermediate points specified for representing a preferred road segment of the highway endpoint along the existing road, for $i = 1, \ldots, n_i$
$\mathbf{I}_{E^{2k-1}}, \mathbf{I}_{E^{2k}}$	A pair of intermediate points corresponding to the selected road segment k for the highway endpoint
I_{PE_k}	A dummy variable indicating if the kth parcel is the environmentally sensitive area
J_A	Delay parameter used in Akcelik's model (1991)
k	The road segment selected for the highway endpoint from the random search process
k_{S_i}	Abscissa of the shifted PC referred to $\mathbf{TS}_i$
L_a	Length of arc a in the highway network; $a \in \mathbf{A}$; L_{a^0}, L_{a^1} = length of arc a^0 and a^1, respectively; $a^0 \in \mathbf{A}^0$ and $a^1 \in \mathbf{A}^1$

Table A.1 (*Continued*)

Notation	Description
L_N	Total length of the highway alignment generated
L_E	Length of a highway section for earthwork volume calculation
L_{PC_i}	Tangent distance from point of circular curve ($\mathbf{PC}_i$) to $\mathbf{PI}_i$
L_{ST_i}	Tangent distance from $\mathbf{PI}_i$ to the endpoint of transition curve ($\mathbf{ST}_i$)
L_{TDE}	Total length of the road segments specified for the endpoint generation
L_{TS_i}	Tangent distance from $\mathbf{TS}_i$ to $\mathbf{PI}_i$;
L_{V_i}	Vertical curve length at the ith vertical curve section
L_{V_m}	Minimum length of vertical curve
l_{B_i}	Length of the ith highway bridge
l_{DE^j}	Length of a road segment, specified for an endpoint generation, for $j = 1, \ldots, n_{seg}$
l_{EP}	Length of roadways (e.g., ramps) associated three-leg structures at the highway endpoint
l_{seg}	Distance from $\mathbf{I}_{E2k-1}$ to the highway endpoint
Δl_{seg}	A provisional distance used for finding the reference points $\mathbf{RP}_1$ and $\mathbf{RP}_2$; (typically less than 10 ft)
l_{STi}	The length of spiral transition curve at i^{th} horizontal curve section
l_T	Tunnel length
l_{Temp}	A provisional random value from $r_c\,[A,\ B]$
$\mathbf{M}_i$	The middle point of the line segment connecting $\mathbf{TS}_i$ to $\mathbf{ST}_i$
m_i	The total number of intersection points of $\overrightarrow{VC_l}$ with property pieces in the *HFB*
m_j	Vehicle maintenance cost of jth mode; $m_{Auto}, m_{Truck} =$ maintenance cost of auto and truck, respectively
$MaxA_k$	Maximum allowable area of the kth land parcel for the new highway construction
N	Node set of a given highway network
N_{ipop}	Total number of sample highways generated in the initial population; $N_{ipop} \geq 5 \times n_{seg1} \times n_{seg2}$
n_B	Total number of bridges in the highway alignment generated
n_E	Total number of highway sections for earthwork volume calculation
n_{HC}	Total number of horizontal curve sections of the highway alignment generated
n_i	Total number of intermediate points specified
n_{OC}	Total number of orthogonal cutting planes where PI's are generated

(*Continued*)

Table A.1 (*Continued*)

Notation	Description
n_{PC}	Total number of land parcels affected by the highway alignment generated
n_{PI}	Total number of PI's that outline the highway alignment generated
n_{seg}	Total number of the road segments specified for endpoint generations; $n_{seg} = n_i/2$
n_{VC}	Total number of vertical curve sections in the highway alignment generated
n_y	Analysis period or design life of a road
o	Average vehicle occupancy, $\mathbf{o} = \left[o_{Auto}, o_{Truck}\right]$; o_{Auto} and o_{Truck} are average vehicle occupancy for autos and trucks, respectively
O_i	The origin of the ith vertical cutting line; $O_i = (x_{O_i}, y_{O_i})$
OC_i	i^{th} orthogonal cutting plane, for $i = 1, \ldots, n$
OB	The origin of the base study area (*BSA*); $OB = (x_{origin}, y_{origin})$
pF_j	Fuel price of jth mode operated on the new highway; pF_{Auto}, pF_{Truck} = fuel prices (\$/gallon) for auto and truck, respectively
ps_i	Offset from the initial tangent to the *PC* of the shifted circle
$\mathbf{PI}_i$	i^{th} PI of the highway alignment generated; $PI_i = (x_i, y_i, z_i)$ for $i = 1, \ldots, n_{PI}$
q_{rs}	Trip rate between origin r and destination s
$r_c[A, B]$	A random value from a continuous uniform distribution whose domain is within the interval $[A, B]$
R_{H_m}	Minimum horizontal curve radius
R_{H_i}	Horizontal curve radius at the ith horizontal curve section
R_T	Tunnel radius
RP	Reference points required to model highway structures at endpoints; $\mathbf{RP}_0 = \left(x_{RP_0}, y_{RP_0}, z_{RP_0}\right)$, $\mathbf{RP}_1 = \left(x_{RP_1}, y_{RP_1}, z_{RP_1}\right)$, $\mathbf{RP}_2 = \left(x_{RP_2}, y_{RP_2}, z_{RP_2}\right)$
$\mathbf{R}(\theta)$	Rotation matrix
r_t	Annual traffic growth rate
S_{a_Auto}, S_{a_Truck}	Average speeds (mile/hr) of autos and trucks on arc a, respectively
S_a^f	Free flow speed of arc a; $a \in A$
s_c, s_f	Cut and fill slops, respectively
S_{H_i}	Horizontal sight distance at the ith horizontal curve section
S_{H_m}	Minimum horizontal sight distance
S_i^l	i^{th} intersection point of $\overrightarrow{VC_l}$ with property pieces that are in the specified horizontal bounds (*HFB*), for $l = 1, \ldots, m_i$
s_r	Earth shrinkage or swell factor
S_{Ti}	Spiral transition curve length at the ith horizontal curve section

Table A.1 (*Continued*)

Notation	Description
S_{T_m}	Minimum length of spiral transition curve
S_{V_i}	Vertical sight distance at the ith vertical curve section
S_{V_m}	Minimum vertical sight distance
$\mathbf{SC}_i$	The point of change from spiral to circle pertaining to $\mathbf{PI}_i$
$\overline{SE}$	A line connecting the *Start* and *End*;
$\mathbf{ST}_i$	The point of change from spiral to tangent pertaining to $\mathbf{PI}_i$
Start, End	Horizontal start and end points of an alignment; *Start* $= (x_s, y_s)$; *End* $= (x_e, y_e)$;
$Start_V$	Start point of a vertical alignment; $Start_V = (H_0, Z_0)$, where H_0 and Z_0 are given
$\mathbf{t}$	A vector of average travel time in different time frames; $\mathbf{t} = [t_{AM}, t_{PM}, t_{OFF}]$
t_a	Travel time on arc a; $\quad \mathbf{t} = (\cdots, t_a \cdots); a \in \mathbf{A}; t_{a^0}^{AM}, t_{a^0}^{PM}, t_{a^0}^{OFF}$ are average travel time (hours) that vehicles spent on arc a^0 during AM, PM, and Off peak hours, respectively before a new highway is added in the network
t_a^f	Free-flow travel time (hr/mile) on arc a; $t_a^f = L_a / V_a^f$
$\mathbf{T}$	Traffic composition vector, $\mathbf{T} = \left[1 - T_{Truck}, T_{Truck} \right]$; T_{Truck} is the proportion of trucks in traffic flow stream over the network
$\dot{T}$	Duration of traffic flow; typically 1 hour
$tempL_i^1$	Provisional lower bound of Z_i based on Z_{i-1}
$tempL_i^2$	Provisional lower bound of Z_i based on Z_{i+1}
$tempU_i^1$	Provisional upper bound of Z_i based on Z_{i-1}
$tempU_i^2$	Provisional upper bound of Z_i based on Z_{i+1}
$\mathbf{TS}_i$	The point of change from tangent to spiral pertaining to $\mathbf{PI}_i$;
U, U'	The area of interest and the area outside interest, respectively
U_k	A dummy variable to indicate whether property piece k is inside U; ($U_k = 1$ if k is inside U; $U_k = 0$, otherwise)
$\mathbf{u}_a$	A vector of unit vehicle operation costs for autos and trucks traveling on arc a
u_A	Unit accident cost
u_c	Unit cut cost
u_E	Unit environmental cost per vehicle mile traveled (VMT)
u_f	Unit fill cost
u_L	Unit length-dependent cost except pavement cost
u_{M^B}	Unit bridge maintenance cost
u_{M^H}	Unit maintenance cost for highway basic segments

(*Continued*)

Table A.1 (*Continued*)

Notation	Description
u_P	Unit pavement cost
u_T	Tunnel earthwork unit cost
u_{v_k}	Unit cost (property value) of the kth land parcel affected by the highway alignment
$\mathbf{v}$	Travel time values for auto and truck, $\mathbf{v} = \left[v_{Auto}, v_{Truck} \right]$; v_{Auto} and v_{Truck} are unit travel time value ($/hr) of auto and truck drivers, respectively
V_i	Vertical feasible gate where VPI_i is generated for $i = 1, \ldots, n_{PI}$
$\overrightarrow{VC_l}$	i^{th} vertical cutting line for i^{th} horizontal PI
VPI_i	i^{th} vertical point of intersection; $VPI_i = (H_i, Z_i)$, for $i = 1, \ldots, n_{PI}$
w_B	Bridge width
w_E	Width of an existing road intersected by the new highway
w_L	Travel lanes width of the new highway
w_N	Width of the new highway; $w_N = w_L + w_S$
w_P	Width of paved portion of the new highway; $w_P \leq w_N$
w_S	Shoulders width of the new highway
$\mathbf{x}$	A vector of average traffic volume in different time frames; $\mathbf{x} = [x_{AM}, x_{PM}, x_{OFF}]$
x_a	Traffic flow on arc a; $\mathbf{x} = (\cdots, x_a, \cdots); a \in A; x_{a^0}^{AM}, x_{a^0}^{PM},$ $x_{a^0}^{OFF}$ are average traffic flows in vehicles per hour on arc a^0 during AM, PM, and Off peak hours, respectively before a new highway is added in the network.
x_{new_CV}	Coefficient of variation of the predicted traffic volumes operating on the initial population members; $x_{new\_CV} = x_{new\_SD}/x_{new\_M}$
x_{new^j}	Predicted traffic volumes that would operate on the jth new highway of the initial population
x_{new_M}, x_{new_SD}	Mean and standard deviation of all x_{new^j}, respectively for $j = 1, \ldots, N_{ipop}$
x_{LB}, y_{LB}, z_{LB}	Lower bounds of x, y, z coordinates of $\mathbf{PI}_i$
x_{UB}, y_{UB}, z_{UB}	Upper bounds of x, y, z coordinates of $\mathbf{PI}_i$
x_{ST_i}	Tangent distance from $\mathbf{TS}_i$ to $\mathbf{SC}_i$ with reference to initial tangent
y_{ST_i}	Tangent offset at $\mathbf{SC}_i$ with reference to $\mathbf{TS}_i$ and initial tangent
x_{CT_i}, y_{CT_i}	Coordinates of ith $\mathbf{CT}$ (circular-to-tangent) point; i.e., end of a circular curve
x_{TC_i}, y_{TC_i}	Coordinates of ith $\mathbf{TC}$ (tangent-to-circular) point; i.e., beginning of a circular curve)

Table A.1 (*Continued*)

Notation	Description
Z_i	Z coordinate at VPI_i, for $i = 1, \ldots n_{PI}$
Z_i^g	Ground elevation at H_i for $i = 1, \ldots n_{PI}$
Z_i^{LB}	Lower bound of Z_i for all $i = 1, \ldots n_{PI}$
Z_i^{UB}	Upper bound of Z_i for all $i = 1, \ldots n_{PI}$
Z_{UL}	Upper level objective function value of the bi-level HAO
Z_{LL}	Lower level objective function value of the bi-level HAO

Appendix B

Traffic Inputs to the HAO Model for the ICC Case Study

Table B.1 TAZ IDs aggregated to hypothetical centroids for ICC case study

Centroid ID	TAZ IDs Aggregated to Centroids	No. of TAZs
1	526~543, 557~576, 605, 607	29
2	512, 516, 517, 530~533, 537~539	10
3	471, 473, 480~483, 487~489, 555, 556, 580	12
4	544~554, 577~579	14
5	467~469, 474, 475, 518, 519, 522~525, 534~536	14
6	376, 387, 388, 471	4
7	407, 470, 476~479, 485	7
8	397, 404~418, 432, 491~493	19
9	381~385, 389~396, 399, 401, 472	15
10	196, 197, 320~328, 336~339, 377~382, 386, 1465~1471	28
11	198~200, 329~331, 340~349, 398, 402, 403, 419	19
12	206, 215~217, 221, 332~335, 347, 350, 351, 420~423, 433, 434	18
13	486, 490, 496, 497, 581, 588, 589	7
14	500, 505, 592	3
15	501~504	4
16	582, 622, 623, 1089	4
17	583~585, 591, 593	5
18	1085~1087	3
19	506~509, 1083, 1084, 1092, 1094, 1095, 1099	10
20	460~466, 781, 783~785, 865~867, 871	15
21	870, 873~875, 877, 878, 1081, 1082, 1091, 1093, 1096~1098	13
22	869, 872, 876, 879, 880, 888~891, 1080, 1090	11
23	884~887, 892~898	11
24	1118, 1119, 1122~1126, 1131, 1133~1136, 1140, 1141	14
25	789~793, 881~883	8
26	670~677	8
27	678~685, 787, 788, 794~815	32
28	218~224, 352~358, 424~431, 440, 477, 478	24
29	235~241, 359~367	16
30	243, 244, 368~374, 451, 454, 455, 642, 648, 650, 655, 656, 659	17
31	375, 456~459, 640, 641, 782	8
32	643~645, 647, 657	5
33	435~446, 452, 453, 494, 498, 499	16

251

Table B.2 AM-peak O/D trips between aggregated centroids for ICC cases study (trips/hr)

O/D	1	2	3	4	5	6	7	8	9	10	11	12	13	14	15	16	17	18	19	20	21	22	23	24	25	26	27	28	29	30	31	32	33	Sum
1	0	1497	639	2751	1095	81	375	756	546	219	387	120	54	9	12	147	75	6	21	39	15	9	3	9	6	6	15	108	108	45	39	6	60	9258
2	1404	0	342	852	885	48	204	384	276	111	198	60	27	3	6	24	18	3	9	18	6	6	3	6	3	3	9	60	54	27	21	3	33	5106
3	636	345	0	1029	513	69	270	450	255	87	159	54	75	9	12	60	51	3	15	21	12	6	3	6	3	3	6	60	45	21	18	3	72	4371
4	2577	798	876	0	642	126	309	534	399	162	255	84	93	15	21	246	138	3	30	33	21	9	3	9	6	3	12	81	75	33	27	6	78	7824
5	765	816	402	642	0	90	378	645	399	156	264	84	42	3	9	24	3	3	12	24	9	3	3	6	3	3	12	84	72	39	30	6	60	5118
6	114	69	90	126	90	0	90	525	255	84	138	39	39	3	3	21	21	3	6	9	6	6	6	9	3	3	6	33	30	18	6	3	15	1563
7	252	147	186	219	81	87	0	738	231	75	138	36	57	3	3	12	21	3	15	15	3	3	3	3	3	3	6	72	30	21	15	3	84	2805
8	474	249	309	369	498	90	738	0	1032	264	735	375	258	24	54	21	63	3	66	138	27	18	3	18	12	6	27	552	192	111	93	12	792	7623
9	429	225	198	342	372	198	297	525	0	708	981	318	39	6	12	21	30	3	30	69	18	15	3	18	12	6	27	258	192	108	78	9	123	6540
10	183	102	75	147	165	129	123	525	708	0	1605	852	12	6	18	9	6	3	18	42	30	24	6	27	18	9	30	168	195	72	48	9	42	4809
11	222	123	96	171	189	60	144	603	876	513	0	318	39	18	18	6	12	3	45	111	30	24	6	24	6	9	42	540	432	162	120	15	150	6798
12	90	45	72	84	114	18	84	600	525	315	1437	0	39	39	18	6	12	3	33	87	33	24	3	12	6	9	33	777	459	138	93	15	270	5193
13	102	51	138	153	114	9	186	57	84	27	108	75	0	39	66	21	78	6	60	87	27	12	3	12	6	3	12	192	57	36	42	6	636	3054
14	15	6	15	24	15	3	21	57	30	6	18	15	39	0	60	9	48	6	63	36	27	12	3	12	6	3	9	33	15	12	18	6	63	675
15	21	12	24	33	27	3	42	153	30	15	63	15	39	54	0	9	90	81	222	192	57	27	3	21	6	3	12	123	57	57	84	6	219	1728
16	333	66	126	483	90	3	39	15	39	15	33	12	72	12	9	0	90	81	123	15	81	27	3	21	3	3	3	18	12	6	6	3	27	1872
17	171	48	120	297	87	3	75	195	39	12	48	30	108	69	51	108	0	18	81	57	54	21	30	30	6	3	6	69	30	18	27	6	123	2007
18	18	3	9	21	6	3	3	15	6	6	6	3	9	3	3	51	21	0	894	597	636	162	3	120	6	3	9	9	9	6	6	6	12	2097
19	36	15	18	42	36	6	33	132	57	30	111	51	45	39	120	72	48	798	0	0	3198	702	21	492	33	12	63	114	105	102	162	24	126	7440
20	36	18	21	30	36	6	39	240	99	51	189	105	54	21	99	6	24	12	495	0	516	303	66	105	252	45	300	261	189	312	480	93	285	4788
21	215	209	12	221	15	3	15	54	33	1021	63	30	18	18	27	33	27	375	2763	372	0	2157	102	1053	108	27	1204	57	63	78	78	54	39	10544
22	209	206	6	209	9	3	9	30	24	15	63	30	6	6	27	33	3	90	495	255	516	0	267	570	36	27	312	36	42	60	57	57	21	5229
23	203	203	3	203	3	3	3	12	39	9	21	9	3	6	3	3	3	6	27	129	294	570	0	201	153	27	333	15	18	30	24	36	3	2586
24	209	206	6	209	12	3	9	39	30	21	57	27	6	6	12	6	12	78	450	135	1599	1011	120	0	108	36	372	48	57	54	48	36	15	5049
25	6	3	3	3	3	3	3	12	6	6	15	9	6	3	3	3	3	3	18	156	99	108	57	39	0	228	231	18	57	42	30	48	15	972
26	3	3	3	3	3	3	3	6	6	6	12	9	3	3	3	3	3	3	6	27	27	27	3	15	24	0	165	12	21	87	30	36	3	624
27	18	9	9	15	21	3	15	69	54	48	96	51	6	15	3	3	6	9	54	459	339	393	183	288	594	228	0	93	96	222	153	258	27	3825
28	99	54	57	78	96	21	99	687	315	171	723	660	81	15	45	6	24	3	75	210	51	39	6	36	30	15	63	0	861	336	216	24	504	5700
29	63	33	30	48	57	15	99	237	168	150	495	327	18	6	18	3	9	3	54	135	54	39	6	42	27	24	63	597	0	444	168	33	87	3492
30	51	30	27	42	60	12	42	267	186	102	312	192	24	9	36	3	9	3	99	435	132	99	18	48	111	114	246	528	444	0	579	231	129	4854
31	36	21	18	27	36	9	27	153	105	54	180	105	21	9	39	3	12	3	120	510	117	87	15	42	81	36	159	255	60	522	0	114	111	3243
32	3	3	3	18	6	3	9	12	9	9	54	3	3	9	6	3	3	3	15	93	54	45	12	18	36	6	180	21	81	174	81	0	12	966
33	87	48	111	105	123	15	207	1410	210	72	330	366	549	51	153	18	75	6	141	330	57	27	6	24	15	6	33	963	183	141	159	12	0	6039
Sum	9080	5666	4017	8933	5937	987	3918	12129	6831	4984	9237	4461	1833	480	957	939	969	1560	6555	4875	9570	6003	945	3339	1905	774	3994	6258	4647	3531	3036	1212	4230	143792

Source: 2010 O/D trips from MWCOG

Table B.3 PM-peak O/D trips between aggregated centroids for ICC case study (trips/hr)

O/D	1	2	3	4	5	6	7	8	9	10	11	12	13	14	15	16	17	18	19	20	21	22	23	24	25	26	27	28	29	30	31	32	33	Sum
1	0	1496	638	2749	1093	79	374	754	544	219	385	119	54	9	10	145	73	5	21	37	15	209	202	208	6	4	15	108	108	45	38	6	59	9827
2	1404	0	341	852	885	46	203	383	274	110	197	60	26	2	5	24	17	1	8	18	5	204	201	204	3	2	8	58	54	25	20	3	32	5675
3	634	343	0	1027	513	69	269	449	255	85	159	53	74	8	12	59	49	2	15	19	10	4	1	6	3	6	6	58	43	21	16	2	70	4336
4	2575	797	876	0	821	60	308	533	398	160	253	83	93	15	19	245	138	8	28	31	19	209	201	209	5	2	10	79	75	31	27	5	76	8390
5	764	816	402	641	0	89	378	645	397	154	263	83	41	1	8	22	2	8	12	7	7	5	6	6	2	1	10	82	71	39	28	4	58	5081
6	112	69	84	88	126	0	80	236	293	126	137	34	6	1	1	3	3	2	4	9	3	2	3	3	2	1	4	32	30	15	11	2	15	1533
7	252	147	184	218	336	33	0	720	230	74	138	60	56	22	12	10	19	4	13	20	6	16	3	18	11	5	5	71	36	21	15	2	82	2778
8	472	247	308	367	498	86	738	0	1030	263	734	373	257	4	54	21	62	4	66	138	25	138	3	18	10	5	27	552	192	111	91	10	792	7593
9	429	225	196	341	370	178	297	1685	0	512	979	265	37	4	10	11	11	1	29	67	17	13	3	18	9	5	26	257	192	107	76	9	122	6502
10	182	102	74	146	164	127	121	525	708	0	1605	318	12	2	4	3	4	1	17	42	1018	13	3	19	9	5	30	167	195	71	48	9	40	5784
11	220	121	96	171	189	60	144	1134	294	1112	0	852	108	5	17	4	10	1	45	30	30	23	5	27	9	8	42	539	430	162	76	14	149	6773
12	90	48	44	71	84	17	84	602	82	314	1437	0	74	7	16	4	10	1	33	86	32	23	2	12	4	1	31	775	136	93	40	13	269	5161
13	101	49	137	153	114	8	185	600	55	26	108	74	0	39	64	19	77	7	59	85	31	12	1	12	4	1	10	191	34	12	17	2	636	3020
14	15	15	23	33	23	2	19	55	14	4	17	13	39	0	59	8	48	7	62	36	31	25	3	12	6	3	3	32	14	57	40	5	63	646
15	21	10	24	33	25	2	40	153	28	13	62	37	64	53	0	7	35	4	220	190	56	21	1	21	6	1	12	122	56	12	83	2	218	1694
16	332	64	126	482	89	2	38	71	37	14	31	11	19	10	0	0	89	79	123	13	79	25	3	20	1	3	3	16	11	5	6	1	26	1835
17	171	47	47	295	86	3	73	193	37	12	46	30	107	68	49	106	0	18	79	56	53	19	1	29	6	6	6	69	29	18	27	4	122	1975
18	16	3	7	19	6	1	6	15	4	2	6	3	7	9	6	51	20	0	892	10	635	19	3	120	4	2	9	8	7	1	6	4	12	2069
19	34	13	23	40	28	4	33	132	57	28	109	51	43	38	119	72	48	798	0	596	3198	701	19	490	33	10	61	114	105	100	162	23	124	7406
20	36	18	18	30	30	3	39	238	31	50	188	103	52	21	97	5	23	10	494	0	516	302	65	105	251	43	299	259	189	310	479	91	285	4758
21	15	7	11	19	13	3	14	53	31	19	61	29	16	16	27	5	25	373	2761	372	0	2157	101	1051	108	27	203	57	61	77	77	54	39	7910
22	8	5	5	9	9	2	7	30	22	15	40	22	6	6	12	8	8	89	495	254	1947	0	267	568	153	35	311	38	41	60	56	56	19	4603
23	3	5	8	8	12	2	3	12	7	7	57	9	2	6	2	1	3	3	27	128	515	267	0	200	170	25	331	15	16	28	22	34	19	1947
24	9	5	8	8	12	2	8	38	30	21	14	27	12	12	21	20	29	120	490	105	1051	568	200	0	38	24	164	11	56	53	47	34	24	4426
25	2	1	2	3	3	1	2	11	9	5	11	7	4	6	6	1	6	4	5	39	108	153	170	38	0	227	63	11	14	87	22	68	5	933
26	17	9	7	13	21	3	13	7	5	5	11	7	1	3	1	3	6	2	24	107	27	35	25	24	227	0	164	92	95	220	153	256	31	585
27	97	54	56	78	94	20	99	68	52	48	96	50	10	3	12	3	6	9	5	0	203	311	331	164	63	23	0	597	677	62	167	23	31	4792
28	61	33	28	47	55	13	37	686	315	169	722	659	191	32	122	16	69	8	75	128	57	38	15	16	11	0	62	0	860	334	215	256	962	5674
29	51	30	25	40	58	12	41	236	167	148	493	326	34	14	56	11	29	7	52	434	61	41	16	53	14	95	63	597	0	220	167	23	183	3457
30	35	19	16	27	36	7	25	184	103	100	311	190	12	57	12	5	18	1	99	135	77	60	28	40	87	220	62	0	522	0	579	113	140	4825
31	51	33	28	40	55	13	37	151	52	52	180	103	17	40	83	6	27	6	118	510	77	85	14	40	81	153	178	527	215	174	0	231	110	3206
32	3	2	2	3	4	2	3	12	7	7	18	11	2	5	2	1	4	3	13	91	54	43	12	18	59	34	178	253	215	81	81	0	110	929
33	85	46	110	105	123	13	206	1409	209	72	365	548	636	63	218	26	122	6	141	330	56	27	24	24	13	31	31	962	183	140	159	16	0	6014
Sum	8250	4835	3984	8102	5911	951	3889	12102	6793	3946	9206	4427	1802	445	921	904	932	1516	6525	4845	11540	6563	1515	3916	1879	739	2964	6223	4620	3504	3011	1178	4199	142137

Source: 2010 O/D trips from MWCOG

Table B.4　OFF-peak O/D trips between aggregated centroids for ICC case study

O/D	1	2	3	4	5	6	7	8	9	10	11	12	13	14	15	16	17	18	19	20	21	22	23	24	25	26	27	28	29	30	31	32	33	Sum
1	0	1496	638	2749	1093	79	374	754	544	219	385	119	54	9	10	145	73	21	21	37	15	9	1	8	6	4	15	108	108	45	38	6	59	9227
2	1404	0	341	852	885	46	203	383	274	110	197	60	26	2	5	24	17	8	8	18	5	4	1	4	3	2	8	58	54	25	20	3	32	5075
3	634	343	0	1027	513	69	269	449	255	85	159	53	74	2	12	59	49	2	15	39	10	4	1	6	3	3	6	58	43	21	16	5	70	4336
4	2575	797	876	0	821	60	308	533	398	160	253	83	93	15	19	245	138	8	28	31	19	5	1	9	4	3	10	79	75	31	27	5	76	7790
5	764	816	402	641	0	89	378	645	397	154	263	83	41	5	8	22	19	1	12	24	7	5	1	6	4	2	4	82	71	39	28	4	58	5081
6	112	69	84	88	126	0	80	236	293	126	137	34	6	7	3	3	19	1	4	9	3	2	1	3	2	1	4	32	30	15	11	2	15	1533
7	252	147	184	218	336	33	0	720	230	74	138	60	56	7	12	10	19	1	13	20	6	3	1	3	2	1	5	71	36	21	15	2	82	2778
8	472	247	308	367	498	86	738	0	1030	263	734	373	257	22	54	21	62	4	66	138	25	16	3	18	11	5	27	552	192	111	91	10	792	7593
9	429	225	196	341	370	178	297	1685	0	512	979	265	37	4	10	11	11	2	29	67	17	13	3	18	10	6	26	257	192	107	76	9	122	6502
10	182	102	74	146	164	127	121	525	708	0	1605	318	12	2	4	3	4	1	17	42	1018	13	1	19	13	6	30	167	195	71	48	9	40	5784
11	220	121	96	171	189	60	144	1134	874	1112	0	852	37	5	17	7	12	1	45	110	30	23	5	27	17	8	42	539	430	162	120	14	149	6773
12	90	48	44	71	84	17	84	602	294	314	1437	0	39	39	16	4	10	1	33	86	32	25	2	23	15	8	31	775	457	136	93	13	269	5161
13	101	49	137	153	114	8	185	600	82	26	108	74	0	0	64	19	77	5	59	85	31	25	2	12	4	2	10	191	56	34	40	2	636	3020
14	15	5	14	23	14	3	19	55	9	4	17	13	39	0	59	8	48	7	62	36	27	12	2	12	4	1	3	32	14	12	17	2	63	646
15	21	10	24	33	25	2	40	153	28	13	62	37	70	53	0	8	35	4	220	190	56	25	3	21	6	3	12	122	56	57	83	5	218	1694
16	332	64	126	482	89	3	38	71	37	14	31	11	24	0	8	0	89	79	123	13	79	21	1	20	1	1	3	16	11	5	6	1	26	1835
17	171	47	119	295	86	3	73	193	37	12	46	30	107	68	49	106	0	18	79	56	53	19	1	29	3	1	6	69	29	18	27	3	122	1975
18	16	13	7	19	6	1	6	15	4	2	6	7	7	9	6	51	20	0	892	21	635	160	120	4	2	9	8	8	7	5	6	4	12	2069
19	34	13	23	40	28	4	33	57	51	103	109	51	43	38	119	72	48	798	0	596	3198	701	19	490	33	10	61	114	189	100	162	23	124	7406
20	36	18	21	30	36	6	39	238	97	50	188	103	52	21	97	5	23	10	494	0	516	302	65	105	251	43	299	259	310	77	54	91	285	7910
21	15	7	11	19	13	3	14	53	31	19	61	16	16	16	27	33	25	373	2761	372	0	2157	101	1051	108	27	203	57	61	77	56	54	39	11540
22	8	5	5	2	8	2	7	30	19	15	40	6	6	6	12	8	8	89	495	254	1947	0	267	568	38	35	311	38	56	41	56	34	19	4603
23	9	3	5	4	12	1	8	12	7	7	19	2	2	6	11	6	10	78	27	128	568	568	0	200	107	36	371	46	16	28	47	34	14	1947
24	2	1	1	2	3	1	2	1	2	5	9	1	2	1	1	1	1	1	27	133	1598	106	120	0	37	24	36	230	16	30	40	30	5	4426
25	9	2	5	4	3	2	8	11	5	5	14	2	6	6	11	1	1	1	16	155	98	26	56	37	0	38	227	16	21	87	34	68	5	933
26	2	1	2	2	1	1	1	5	2	2	5	2	2	1	1	1	1	2	5	39	27	26	56	14	38	0	164	11	21	34	6	3	6	585
27	17	7	9	13	21	13	5	68	48	96	722	659	79	14	45	6	3	5	54	457	1337	391	183	288	593	227	0	92	860	220	153	256	26	4792
28	97	54	56	78	94	20	99	686	315	169	722	493	18	5	17	6	23	5	75	209	51	38	6	35	30	14	63	0	41	334	215	23	503	5674
29	61	33	28	47	55	13	37	236	167	148	493	326	18	5	17	2	8	2	52	135	52	37	6	40	27	23	62	597	0	444	167	33	86	3457
30	51	30	25	40	58	12	41	266	184	100	311	190	22	8	34	3	9	2	99	434	132	97	18	48	109	114	245	527	444	0	522	231	129	4825
31	35	19	16	27	36	7	25	151	103	52	180	103	21	8	38	3	11	1	118	510	115	85	14	81	81	34	158	253	677	522	0	579	110	3205
32	3	2	3	2	4	3	3	12	9	9	18	11	2	1	3	1	2	3	13	91	52	43	4	18	59	64	178	31	215	174	81	0	10	923
33	85	46	110	105	123	13	206	1409	209	72	330	365	548	51	153	16	75	141	330	330	56	27	4	24	13	6	31	962	183	140	159	16	0	6014
Sum	8250	4835	3984	8102	5911	951	3889	12102	6793	3946	9206	4427	1802	445	921	904	932	1516	6525	4845	11540	5963	915	3316	1879	739	2964	6223	4620	3504	3011	1178	4199	140337

Source: 2010 O/D trips from MWCOG

References

AASHTO (1997), *"The Value of Travel Time: Departmental Guidance for Conducting Economic Evaluations"*, American Association of State Highway and Transportation Officials, Washington, D.C.

AASHTO (2003), *"User Benefit Analysis for Highways"*, American Association of State Highway Administration and Transportation Officials, Washington, D.C.

AASHTO (2010), *"User and Non-user Benefit Analysis for Highways"*, American Association of State Highway Administration and Transportation Officials, Washington, D.C.

AASHTO (2011), *"A Policy on Geometric Design of Highways and Streets"*, American Association of State Highway and Transportation Officials, Washington, D.C.

Adeli, H. and Cheng, N. T. (1993), "Integrated Genetic Algorithm for Optimization of Space Structures", *Journal of Aerospace Engineering, ASCE*, 6(4), pp. 315–28.

Akcelik, R. (1981), *"Traffic Signals: Capacity and Timing Analysis"*, Research Report ARR 123, Australian Road Research Board.

Akcelik, R. (1991), "Travel Time Functions for Transport Planning Purposes: Davidson's Function, Its Time-Dependent Form and an Alternative Travel Time Function", *Australian Road Research*, 21(3), pp. 49–59.

American Automobile Association (1999), "Your Driving Costs", *AAA Association Communication*, FL, USA.

Athanassoulis, G. C. and Calogero, V. (1973), *"Optimal Location of a New Highway from A to B: A Computer Technique for Route Planning"*, PTRC Seminar Proceedings on Cost Models and Optimization in Highways (Session L9), London.

Babapour, R., Naghdi, R., Ghajar, I., and Mortazavi, Z. (2018), "Forest Road Profile Optimization Using Meta-Heuristic Techniques", *Applied Soft Computing*, 64, pp. 126–137. https://doi.org/10.1016/j.asoc.2017.12.015.

Back, T., Hoffmeister, F. and Schwefel, H.-P. (1991), "A Survey of Evolution Strategies", *Proceedings of Fourth International Conference on Genetic Algorithms*, Morgan Kaufmann Publishers, pp. 2–9.

Back, T. and Khuri, S. (1994), "An Evolutionary Heuristic for the Maximum Independent Set Problem", *Proceedings of the First IEEE Conference on Evolutionary Computation*, IEEE Press, pp. 531–535.

Back, T. (1996), *Evolutionary Algorithms in Theory and Practice: Evolution Strategies, Evolutionary Programming, Genetic Algorithms*, Oxford University Press, New York, Oxford.

Bassani, M., Marinelli, G., and Piras, M. (2016), "Identification of Horizontal Circular Arc from Spatial Data Sources", *Journal of Surveying Engineering*, ASCE, 14(4), https://doi.org/10.1061/(ASCE)SU.1943-5428.0000186.

Bauer, K. M. and Harwood, D. W. (2000), "*Statistical Models of At-Grade Intersection Accidents-Addendum*", FHWA-RD-99-094, Federal Highway Administration, U.S. Department of Transportation.

Beckman, M. J., McGuire, C. B., and Winsten, C. B. (1956), *Studies in the Economics of Transportation*, Yale University Press, New Haven, Connecticut.

Bonabeau, E., Theraulaz, G., and Dorigo, M. (1999). *Swarm Intelligence: From Natural to Artificial Systems*, Oxford University Press.

Bosurgi, G., Pellegrino, O. and Sollazzo, G. (2013), "A PSO Highway Alignment Optimization Algorithm Considering Environmental Constraints", *Advances in Transportation Studies*, 31, pp. 63–80.

Bosurgi, G., Pellegrino, O. and Sollazzo, G. (2014), "Using Genetic Algorithms for Optimizing the PPC in the Highway Horizontal Alignment Design", *Journal of Computing in Civil Engineering*, 30(1). doi.org/10.1061/(ASCE)CP.1943-5487.0000452

Bosurgi, G. Pellegrino, O. and Sollazzo, G. (2017), "A PSO Algorithm for Designing 3D Highway Alignments Adopting Polynomial Solutions, *Proceedings of the AIIT International Congress on Transport Infrastructure and Systems*, Taylor & Francis Group, pp. 385–393, https://doi.org/10.1201/9781315281896.

Bruynooghe, M. (1972), "An Optimal Method of Choice of Investments in a Transport Network", *Planning & Transport Research & Computation (PTRC) Seminar Proceedings on Urban Traffic Model Research*, London, UK.

Bureau of Public Roads (BPR), (1964), "*Traffic assignment manual*", Urban Planning Division, U.S. Department of Commerce, Washington, D.C.

Carlson, S. E. and Shonkwiler, R. (1998), "Annealing a Genetic Algorithm Over Constraints", *Proceedings of IEEE international Conference on Systems, Man, and Cybernetics*, San Diego, CA, USA. DOI:10.1109/ICSMC.1998.726702.

Chan, W. T. and Tao, F. (2003), "Using GIS and Genetic Algorithm in Highway-Alignment Optimization", *Proceedings of Intelligent Transportation Systems*, IEEE Press, pp. 1563–1567.

Chapra, S. C. and Canale, R. P. (1988), *Numerical Methods for Engineers*, McGraw-Hill, Inc., New York.

Chatterjee, A., Everett, J. D., Reiff, B., Schwetz, T. B., Seaver, W. L., and Wegmann, F. J. (2003), "*Tools for Assessing Safety Impact of Long-Range Transportation Plans in Urban Areas*", Federal Highway Administration, U.S. Department of Transportation.

Chen, M. and Alfa, A. S. (1991), "A Network Design Algorithm using a Stochastic Incremental Traffic Assignment Approach", *Transportation Science*, 25(3), pp. 215–224, doi.org/10.1287/trsc.25.3.215.

Chen, A. and Yang, C. (2004), "*Stochastic Transportation Network De*sign Problem with Spatial Equity Constraint", *Transportation Research*, Record No. 1882, pp. 97–104.

Chen, C. C., Wang, S. L. and Schonfeld, P. (2010), "Dynamic Path Optimization for Robotic Agents," *Proceedings of Transportation Research Board, 90th Annual Meeting*, TRB ID: 10-1420, Washington, DC.

Chen, C. C., Hwang, D., Wang, S. L. and Schonfeld, P. (2011), "Incorporating GIS-Based Visibility Analysis into a Military Path Planning Algorithm," *Proceedings of Transportation Research Board*, 91st Annual Meeting, TRB ID: 11-2253, Washington, DC.

Cheng, J. F. and Lee, Y. (2006), "Model for Three-Dimensional Highway Alignment", *Journal of Transportation Engineering*, 132(12), pp. 913–920.

Chew, E. P., Goh, C. J., and Fwa, T. F. (1989), "Simultaneous Optimization of Horizontal and Vertical Alignments for Highways", *Transportation Research Part B*, 23(5), pp. 315–329.

Christian, J. and Newton, L. (1999), "Highway Construction and Maintenance Costs", *Canadian Journal of Civil Engineering*, 26, pp. 445–452.

Clifton, K. J. and Mahmassani, H. S. (2004), "*Economic Impact Study of the Intercounty Connector*", Research Report to Maryland State Highway Administration, Maryland Transportation Initiative, University of Maryland.

Coello, C. A. C. (2000), "Use of a Self-adaptive Penalty Approach for Engineering Optimization Problems", *Computers in Industry*, 41, pp. 113–127.

Coello, C. A. C. (2002), "Theoretical and Numerical Constraint Handling Techniques used with Evolutionary Algorithms: A Survey of the State of the Art", *Computer Methods in Applied Mechanics and Engineering*, 191(11-12), pp. 1245–1287.

Craenen, B. G. W., Eiben, A. E., and Marchiori, E. (2001), "How to Handle Constraints with Evolutionary Algorithms", in *The Practical Handbook of Genetic Algorithms, Applications*, Edited by Lance Chamers, Chapter 10, pp. 341–361, Chapman & Hall/CRC.

Craenen, B. G. W., Eiben, A. E., and van Hemert, J. I. (2003), "Comparing Evolutionary Algorithms on Binary Constraint Satisfaction Problems", *IEEE Transactions on Evolutionary Computation*, 7(5), pp. 424–444.

Dasgupta, D. and Michalewicz, Z. (1997), *Evolutionary Algorithms in Engineering Applications*, Springer-Verlag, Berlin.

Davey, Dunstall, and Halgamuge (2017), "Optimal Road Design Through Ecologically Sensitive Areas Considering Animal Migration Dynamics", *Transportation Research Part C: Emerging Technologies*, 77 (April 2017), pp. 478–494.

Davidson, K. B. (1966), "A Flow-Travel Time Relationship for Use in Transportation Planning", *Proceedings of the Third APRB Conference*, 3(1), pp. 183–194.

Davidson, K. B. (1978), "The Theoretical Basis of a Flow-Travel Time Relationship for Use in Transportation Planning", *Australian Road Research*, 8(1), pp. 32–35; Discussion, pp. 45.

Davis, L. (1991), *Handbook of Genetic Algorithms*, Van Nostrand Reinhold, New York.

Davis, G. A. (1994), "Exact Local Solution of the Continuous Network Design Problem via Stochastic User Equilibrium Assignment", *Transportation Research Part B*, 28(1), pp. 61–75.

de Smith, M. J. (2006), "Determination of Gradient and Curvature Constrained Optimal Paths", *Computer-Aided Civil and Infrastructure Engineering*, 21, pp. 24–38.

Dowling, R. G., Kittelson, W., Skabardonis, A., and Zegeer, J. (1997), "*Techniques for Estimating Speed and Service Volumes for Planning Applications*", NCHRP Report 387, Transportation Research Board, National Research Council.

Dowling, R. G., Singh, R., and Cheng, W. W. K. (1998), *"The Accuracy and Performance of Improved Speed-Flow Curves"*, Transportation Research, Record No. 1646, pp. 9–17.

Easa, S. M. (1988), "Selection of Roadway Grades that Minimize Earthwork Cost Using Linear Programming", *Transportation Research Part A*, 22(2), 121–136.

Easa, S. M., Strauss, T. R., Hassan, Y. and Souleyrette, R. R. (2002), "Three-dimensional Transportation Analysis: Planning and Design", *Journal of Transportation Engineering*, 128(3), pp. 250–258.

Easa, S. M. and Mehmood, A. (2008), "Optimizing Design of Highway Horizontal Alignments: New Substantive Safety Approach", *Computer-Aided Civil and Infrastructure Engineering*, 23(7), pp. 560–573.

Eccleston, C. (2008), *NEPA and Environmental Planning: Tools, Techniques, and Approaches for Practitioners*, CRC Press.

Eiben A. E. and van der Hauw J. K. (1998), "Adaptive Penalties for Evolutionary Graph Coloring," In *Artificial Evolution: AE 1997*, Hao J.K., Lutton E., Ronald E., Schoenauer M., Snyers D. (eds.), Lecture Notes in Computer Science, Vol. 1363, Springer, Berlin, Heidelberg, https://doi.org/10.1007/BFb0026593.

Engelbrecht, A. (2007), *"Computational Intelligence: An Introduction"*, Wiley.

FHWA (2000), *"Surface Transportation Efficiency Analysis Model (STEAM 2.0): User Manual"*, Federal Highway Administration, Washington, DC.

FHWA (2002), *"Highway Economic Requirements System State Version (HERS-ST): User's Guide"*, Federal Highway Administration, Washington, D.C.

FHWA and MDSH (1997), "Draft Environmental Impact Statement and Major Investment Study", in *Volume 1: Evaluation of Intercounty Connect from I-270 to US* 1. US Department of Transportation.

Floudas, C.A., Pardalos, P., Adjiman, C., Esposito, W. R., Gumus, Z. H., Harding, S. T., Klepeis, J. L., Meyer, C. A., and Schweiger, C. A. (1999), *Handbook of Test Problem in Local and Global Optimization*, Springer US.

Friesz, T., Cho, H.-J., Mehta, N., Tobin, R., and Anandalingam, G. (1992), "A Simulated Annealing Approach to the Network Design Problem with Variational Inequality Constraints", *Transportation Science*, 26(1), pp. 18–26.

Fwa, T. F. (1989), *"Highway Vertical Alignment Analysis by Dynamic Programming"*, Transportation Research, Record No. 1239, pp. 1–9.

Fwa, T. F., Chan, W. T., and Sim, Y. P. (2002), "Optimal Vertical Alignment Analysis for Highway Design", *Journal of Transportation Engineering*, 128(5), pp. 395–402.

Gao, Z., Wu, J., and Sun, J. (2005), "Solution Algorithm for the Bi-level Discrete Network Design Problem", *Transportation Research Part B*, 39, pp. 479–495.

Garber, N. J. and Hoel, L. A. (1998), *Traffic and Highway Engineering*, 2nd edition, PWS Publishing, New York.

Gen, M. and Cheng, R. (1996), "A Survey of Penalty Techniques in Genetic Algorithms", *Proceedings of the Third IEEE Conference on Evolutionary Computation*, IEEE Press, pp. 804–809.

Ghoreishi, B., Shafahi, Y., and Hashemian, S. E. (2019), "A Model for Optimizing Railway Alignment by Considering Costs of Bridge and Tunnel and Using the Transitional Curves", *Urban Rail Transit*, pp. 1–18. doi.org/10.1007/s40864-019-00111-5

Glennon, J., Newman, T., and Leisch, J. (1985), *Safety and Operational Considerations for Design of Rural Curves.* FHWA/RD-86/035, Federal Highway Administration, Washington, DC.

Glover, F. (1977), "Heuristics for Integer Programming Using Surrogate Constraints", *Decision Sciences*, 8(1), pp. 156–166.

Glover, F. and Kochenberger, G. (1995), "Critical Event Tabu Search for Multi-dimensional Knapsack Problems", *Proceedings of the International Conference on Metaheuristics for Optimization*, Kluwer Publishing, pp. 113–133.

Goh C. J., Chew, E. P., and Fwa, T. F. (1988), "Discrete and Continuous Models for Computation of Optimal Vertical Highway Alignment", *Transportation Research Part B*, 22(6), pp. 399–409.

Goldberg, D. E. (1989), *Genetic Algorithms in Search, Optimization, and Machine Learning*, Addison-Wesley Publishing Company, Inc., Massachusetts.

Hadj-Alouane, A. B. and Bean, J. C. (1992), "A Genetic Algorithm for the Multiple-choice Integer Program", *Operations Research*, 45, pp. 92–101.

Haling, D. and Cohen, H. (1996). "Residential Noise Damage Costs Caused by Motor Vehicles", *Transportation Research Record*, 1559(1), pp. 84–93. https://doi.org/10.1177/0361198196155900111.

Hare, W., Lucet, Y., and Rahman, F. (2015), "A Mixed-integer Linear Programming Model to Optimize the Vertical Alignment Considering Blocks and Side-slopes in Road Construction", *European Journal of Operational Research*, 241, pp. 631–641.

Halvorsen, R. and Ruby, M. (1981). "Benefit-Cost Analysis of Air-Pollution Control", Lexington MA: Lexington Books, D.C. Heath Co., 1981, pp. 264.

Harwood, D. W., Council, F. M., Hauer, E., Hughes, W. E., and Vogt, A. (2000), *"Prediction of the Expected Safety Performance of Rural Two-Lane Highways"*, FHWA-RD-99-207, Federal Highway Administration, U.S. Department of Transportation.

Hasany, R.M. and Shafahi, Y. (2017), "Ant Colony Optimisation for Finding the Optimal Railroad Path", *Proceedings of the Institution of Civil Engineers — Transport*, 170(4), pp. 218–230.

Hayman, R. W. (1970), *"Optimization of Vertical Alignment for Highways Through Mathematical Programming"*, Highway Research Record No. 306, pp. 1–9.

Hickerson, T. F. (1964), *Route Location and Design*, The 5th Edition, McGraw-Hill, Inc., New York.

Higuera de Frutos, S. and Castro, M. (2017), "A Method to Identify and Classify the Vertical Alignment of Existing Roads", *Computer-Aided Civil and Infrastructure Engineering*, 32, pp. 952–963.

Hirpa, D., Hare, W., Lucet, Y., Pushak, Y., and Tesfamariam, S. (2016), "A Bi-Objective Optimization Framework for Three-Dimensional Road Alignment Design", *Transportation Research Part C*, 65, pp. 61–78.

Hogan, J. D. (1973), "Experience with OPTLOC: Optimum Location of Highway by Computer", *PTRC Seminar Proceedings on Cost Models and Optimization in Highways* (Session L10), London.

Homaifar, A., Lai, S. H.-Y., and Qi, X. (1994), "Constrained Optimization via Genetic Algorithms", *Simulation*, 62, pp. 242–254.

Howard, B. E., Bramnick, Z., and Shaw, J. F. B. (1968), "Optimum Curvature Principle in Highway Routing", *Journal of the Highway Division*, 94(HW1), pp. 61–82.

Huang, W.-C., Kao, C.-Y., Horng, J.-T. (1994), "A Genetic Algorithm Approach for Set Covering Problem", *Proceedings of the First IEEE Conference on Evolutionary Computation*, IEEE Press, pp. 569–573.

Jha, M. K. (2000), "*A Geographic Information Systems-Based Model for Highway Design Optimization*", Ph.D. Dissertation, University of Maryland, College Park.

Jha, M. K. and Schonfeld, P. (2000a), "Integrating Genetic Algorithms and GIS to Optimize Highway Alignments", *Transportation Research*, Record No. 1719, pp. 233–240.

Jha, M. K. and Schonfeld, P. (2000b), "Geographic Information System-Based Analysis of Right-of-Way Cost for Highway Optimization", *Transportation Research*, Record No. 1719, pp. 241–249.

Jha, M. and Schonfeld, P. (2003), "Tradeoffs between Initial and Maintenance Costs of Highways in Cross – slopes", *Journal of Infrastructure Systems*, 9(1), pp. 16–25.

Jha, M. K. and Schonfeld, P. (2004), "A Highway Alignment Optimization Model using Geographic Information Systems", *Transportation Research Part A*, 38(6), pp. 455–481.

Jha, M. K., Schonfeld, P., Jong, J.-C., and Kim, E. (2006), *Intelligent Road Design*, WIT Press, Southhampton, UK.

Jha, M. K. and Kim, E. (2006), "Highway Route Optimization Based on Accessibility, Proximity, and Land-Use Changes", *Journal of Transportation Engineering*, 132(5), https://doi.org/10.1061/(ASCE)0733-947X(2006)132:5(435).

Jha, M., Schonfeld, P., and Samanta, S. (2007), "Optimizing Rail Transit Routes with Genetic Algorithms and GIS", *Journal of Urban Planning and Development*, 133(3), pp. 161–171.

Jha, M. K. and Maji, A. (2007), "A Multi-Objective Genetic Algorithm for Optimizing Highway Alignments", in *Proceedings of IEEE Symposium on Computational Intelligence in Multi-Criteria Decision-Making*, Honolulu, HI, pp. 261–266.

Jha, M. K., Chen, C. C., Schonfeld, P., and Kikuchi, S. (2008), "An Evolutionary Algorithm for Military Applications", *Proceedings of IEEE 3rd International Conference on Systems of Systems Engineering* (SoSE 2008), Monterey, CA, June 2–4, 2008.

Jha, M. K. and Kang, M.-W. (2009), "GIS-Based Model for Highway Noise Analysis", *Journal of Infrastructure Systems*, 15(2), pp. 88–94.

Johnson, D. S., Lenstra, J. K., and Rinooy Kan, A. H. G. (1978), "The Complexity of the Network Design Problem", *Networks*, 8, pp. 279–285.

Joines, J. A. and Houck, C. R. (1994), "On the use of Non-stationary Penalty Functions to Solve Nonlinear Constrained Optimization Problems with GA's", *Proceedings of the First IEEE Conference on Evolutionary Computation*, pp. 579–584.

Jong, J.-C. (1998), "*Optimizing Highway Alignments with Genetic Algorithms*", Ph.D. Dissertation, University of Maryland, College Park.

Jong, J.-C. and Schonfeld, P. (1999), "Cost Functions for Optimizing Highway Alignments", *Transportation Research*, Record No. 1659, pp. 58–67.

Jong, J.-C., Jha, M. K., and Schonfeld, P. (2000), "Preliminary Highway Design with Genetic Algorithms and Geographic Information Systems", *Computer-Aided Civil and Infrastructure Engineering*, 15(4), pp. 261–271.

Jong, J.-C. and Schonfeld, P. (2003), "An Evolutionary Model for Simultaneously Optimizing 3-Dimensional Highway Alignments", *Transportation Research Part B*, 37(2), pp. 107–128.

Kang, M. W., Jha, M. K., and Schonfeld, P. (2005), *"3D Highway Alignment Optimization for Brookeville Bypass"*, Research Report to Maryland State Highway Administration, University of Maryland.

Kang, M. W., Jha, M. K., and Schonfeld, P. (2006), "Three-Dimensional Highway Alignment Optimization for Brookeville Bypass", *Proceedings of Transportation Research Board*, 86th Annual Meeting, TRB ID: 06-1023, Washington, DC.

Kang, M. W., Schonfeld, P., Jha, M. K., and Gautham, K. K. (2007a), *"Improved Alignment Optimization and Evaluation"*, Research Report to Maryland State Highway Administration, University of Maryland.

Kang, M. W., Schonfeld, P., and Jong, J.-C. (2007b), "Highway Alignment Optimization through Feasible Gates", *Journal of Advanced Transportation*, 41(2), pp. 115–144.

Kang, M.-W. (2008) *"An Alignment Optimization Model for A Simple Highway Network"*. Ph.D. Dissertation, University of Maryland, College Park, May 2008.

Kang, M.-W., Schonfeld, P., and Yang, N. (2009), "Prescreening and Repairing in a Genetic Algorithm for Highway Alignment Optimization", *Computer-Aided Civil and Infrastructure Engineering*, 24(2), pp. 109–119.

Kang, M.-W., Yang, N., Schonfeld, P., and Jha, M. K. (2010), "Bilevel Highway Route Optimization", *Transportation Research Record: Journal of the Transportation Research Board*, 2197, pp. 107–117.

Kang, M.-W., Jha, M. K., and Hwang, D. (2011), "A GIS-Based Simulation Model for Military Path Planning of Unmanned Ground Robots", *International Journal of Safety and Security Engineering*, 1(3), pp. 248–264.

Kang, M.-W., Jha, M. K., and Schonfeld, P. (2012), "Applicability of Highway Alignment Optimization Models", *Transportation Research Part C*, 21(1), pp. 257–286.

Kang, M.-W. Shariat, S., and Jha, M. K. (2013), "New Highway Geometric Design Methods for Minimizing Vehicular Fuel Consumption and Improving Safety", *Transportation Research Part C*, 31, pp. 99–111, https://doi.org/10.1016/j.trc.2013.03.002.

Kang, M.-W., Jha, M. K., and Buddharaju, R. (2014), "A Rail Transit Route Optimization Model for Rail Infrastructure Planning and Design: Case Study of St Andrews, Scotland", *Journal of Transportation Engineering*, 140(1), pp. 1–11.

Kazarlis, S. and Petridis, V. (1998), "Varying Fitness Functions in Genetic Algorithms: Studying the Rate of Increase in the Dynamic Penalty Terms", *Proceedings of the Fifth International Conference on Parallel Problem Solving from Nature,* Springer-Verlag, pp. 211–220.

Kennedy, J. and Eberhart, R. (1995), "Particle Swarm Optimization", *Proceedings of IEEE International Conference on Neural Networks*, pp. 1942–1948.

Khan, S., Shanmugam, R., and Hoeschen, B. (1999), "Injury, Fatal, and Property Damage Accident Models for Highway Corridors", *Transportation Research*, Record No. 1665, pp. 84–92.

Kim, D. G. and Husbands, P. (1997), *"Riemann Mapping Constraint Handling Method for Genetic Algorithms"*, Technical Report CSRP 469, COGS, University of Sussex, UK.

Kim, D. G. and Husbands, P. (1998), "Mapping Based Constraint Handling for Evolutionary Search: Thurston's Circle Packing and Grid Generation", in *The Integration of*

Evolutionary and Adaptive Computing Technologies with Product/System Design and Realization, edited by Ian Parmee, Springer-Verlag, pp. 161–173.

Kim, D. N. and Schonfeld, P. (1997), "Benefits of Dipped Vertical Alignments for Rail Transit Routes," *Journal of Transportation Engineering*, 123(1), pp. 20–27.

Kim, E. (2001), "*Modeling Intersections & Other Structures in Highway Alignment Optimization*", Ph.D. Dissertation, University of Maryland, College Park.

Kim, E., Jha, M. and Schonfeld, P. (2004a), "Intersection Construction Cost Functions for Alignment Optimization", *Journal of Transport Engineering*, 130(2), pp. 194–203.

Kim, E., Jha, M. K., Lovell, D. J., and Schonfeld, P. (2004b), "Intersection Cost Modeling for Highway Alignment Optimization", *Computer-Aided Civil and Infrastructure Engineering*, 19(2), pp. 136–146.

Kim, E., Jha, M. K., and Son, B. (2005), "Improving the Computational Efficiency of Highway Alignment Optimization Models through a Stepwise Genetic Algorithms Approach," *Transportation Research Part B*, 39(4), pp. 339–360.

Kim, E., Jha, M. K., Schonfeld, P., and Kim, H. (2007), "Highway Alignment Optimization Incorporating Bridges and Tunnels", *Journal of Transportation Engineering*, 133(2), pp. 71–81.

Kim, M., Markovic, N. and Kim, E. (2019), "A Vertical Railroad Alignment Design with Construction and Operating Costs", *Journal of Transportation Engineering, Part A: Systems*, 145(10), 04019043, https://doi.org/10.1061/JTEPBS.0000269.

Kowalczyk, R. (1997), "Constraint Consistent Genetic Algorithms", *Proceedings of the Fourth IEEE Conference on Evolutionary Computation*, IEEE Press, pp. 343–348.

Koziel, S. and Michalewicz, Z. (1998), "A Decoder-based Evolutionary Algorithm for Constrained Parameter Optimization Problems", *Proceedings of the Fifth Parallel Problem Solving from Nature (PPSNV)*, Springer-Verlag, pp. 231–240.

Kyte, C. A., Perfater, M. A., Haynes, S., and Lee, H. W. (2003), "*Developing and Validating a Highway Construction Project Cost Estimation Tool*", Research Report, Virginia Transportation Research Council.

Lai, X. (2012), "*Optimization of Station Locations and Track Alignments for Rail Transit Lines,*" *for Rail Transit Lines*", Ph.D. Dissertation, University of Maryland, College Park, May 2012.

Lai, X. and Schonfeld, P. (2012), "Optimization of Rail Transit Alignments Considering Vehicle Dynamics", *Transportation Research*, Record No. 2275, pp. 77–87.

Lai, X. and Schonfeld, P. (2016), "Concurrent Optimization of Rail Transit Alignment and Stations", *Urban Rail Transit*, 2(1), pp. 1–15.

Lau, M. Y.-K. and May, Jr. A. (1988), "Injury Accident Prediction Models for Signalized Intersections", *Transportation Research,* Record No. 1172, pp. 58–67.

LeBlanc, L. J. (1975), "An Algorithm for the Discrete Network Design Problem", *Transportation Science*, 9, pp. 183–199.

Lee, Y., Tsou, Y.-R., and Liu, H.-L. (2009), "Optimization Method for Highway Horizontal Alignment Design", *Journal of Transportation Engineering*, 135(4), pp. 217–224.

Le Riche, R. G. and Haftka, R. T. (1994), "Improved Genetic Algorithm for Minimum Thickness Composite Laminate Design", *Composites Engineering*, 3(1), pp. 121–139.

Le Riche, R., Vayssade, C., and Haftka, R. T. (1995), "A Segregated Genetic Algorithm for Constrained Optimization in Structural Mechanics", *Proceedings of the*

Sixth International Conference on Genetic Algorithms, Morgan Kaufmann Publishers, pp. 558–565.

Li, W., L., Pu, H., Schonfeld, P., Zhang, H., and Zheng, X. (2016), "Methodology for Optimizing Constrained 3-dimensional Railway Alignments in Mountainous Terrain," *Transportation Research Part C: Emerging Technologies*, 68, pp. 549–565.

Li, W., Pu, H., Schonfeld, P., Yang, J., Zhang, H., Wang, L., and Xiong, J. (2017), "Mountain Railway Alignment Optimization with Bidirectional Distance Transform and Genetic Algorithm", *Computer-Aided Civil and Infrastructure Engineering*, 32, pp. 691–709.

Li, W., Pu, H., Schonfeld, P., Song, Z., Zhang, H., Wang, L., Wang, J., Peng, X. and Peng, J. (2019), "A Method for Automatically Re-creating the Horizontal Alignment Geometry of Existing Railways," *Computer-Aided Civil and Infrastructure Engineering*, 34, pp 71–94.

Liepins, G. E. and Vose, M. D. (1990), "Representational Issues in Genetic Optimization", *Journal of Experimental and Theoretical Computer Science*, 2(2), pp. 4–30.

Liepins, G. E. and Potter, W. D. (1991), "A Genetic Algorithm Approach to Multiple Fault Diagnosis", in *Handbook of Genetic Algorithms*, edited by Lawrence Davis, Chapter 17, Van Nostrand Reinhold, New York, pp. 237–250.

Lo, H. K. and Tung, Y. K. (2001), "Network Design for Improving Trip Time Reliability", *Proceedings of Transportation Research Board*, 80th Annual Meeting, TRB ID: 01-2204.

Lovell, D. J. (1999), "Automated Calculation of Sight Distance from Horizontal Geometry", *Journal of Transportation Engineering*, 125(4), pp. 297–304.

Lovell, D. J., Jong, J. C. and Chang, P. C. (2001), "Improvements to Sight Distance Algorithm", *Journal of Transportation Engineering*, 127(4), pp. 283–288

Magnanti, T. L. and Wong, R. T. (1984), "Network Design and Transportation Planning: Models and Algorithms", *Transportation Science*, 18, pp. 1–55.

Maji, A. and Jha, M. K. (2009), "Multi-Objective Highway Alignment Optimization Using a Genetic Algorithm", *Journal of Advanced Transportation*, 43(4), pp. 481–504.

Maji, A. and Jha, M. K. (2011), "A Multiobjective Analysis of Impacted Area of Environmentally Preserved Land and Alignment Cost for Sustainable Highway Infrastructure Design", *Procedia – Social and Behavioral Science*, 20, pp. 966–972.

Maji, A. and Jha, M. K. (2012), "Comparison of Single and Multi-Objective Highway Alignment Optimization Algorithms", *Advances in Transportation Studies*, 27, pp. 5–16.

Manheim, M. L. (1970), *Fundamentals of Transportation Systems Analysis, Volume 1: Basic Concepts*, MIT Press.

(MDP) Maryland Department of Planning. "Property Map Products", *MdProperty View Advanced Desktop GIS to use with ESRI's ArcGIS Software*. Available at: https://planning.maryland.gov/Pages/OurProdu cts/PropertyMapProducts/MDPropertyViewProducts.aspx (accessed on October 1, 2018).

MDSHA (1999), "*Consolidated Transportation Program*", Maryland State Highway Administration.

MDSHA (2013), "MD 97 Brookeville Project Smart Growth Package", Maryland State Highway Administration, Baltimore, Maryland.

Meng, Q., Lee, D. H., Yang, H., and Huang, H. J. (2004), "Transportation Network Optimization Problems with Stochastic User Equilibrium Constraints", *Transportation Research*, Record No. 1882, pp. 113–119.

Menn, C. (1990), *Prestressed Concrete Bridges*, Springer-Verlag, Wien.

Meyer, M. D. and Miller, E. J. (2001), *Urban Transportation Planning,* McGraw-Hill, Inc., New York.

Meyer, M. (2016), "Chapter 4: Environmental and Energy Considerations", in *Transportation Planning Handbook, 4th Edition, Institute of Transportation Engineers*, pp. 117–160, Wiley.

Michalewicz, Z. and Janikow, C. Z. (1991), "Handling Constraints in Genetic Algorithms", *Proceedings of the Fourth International Conference on Genetic Algorithms*, Morgan Kaufmann Publishers, pp. 151–157.

Michalewicz, Z., Vignaux, G. A., and Hobbs, M. (1991), "A Non-Standard Genetic Algorithm for the Nonlinear Transportation Problem", *ORSA Journal on Computing*, 3(4), pp. 307–316.

Michalewicz, Z. and Attia, N. (1994), "Evolutionary Optimization of Constrained Problems", *Proceedings of the Third Annual conference on Evolutionary Programming*, World Scientific, pp. 98–108.

Michalewicz, Z. (1995), "Genetic Algorithms, Numerical Optimization, and Constraints", *Proceedings of the Sixth International Conference on Genetic Algorithms*, Morgan Kaufmann Publishers, pp. 151–158.

Michalewicz, Z. and Michalewicz, M. (1995), "Pro-Life Versus Pro-Choice Strategies in Evolutionary Computation Techniques", *Computational Intelligence: A Dynamic System Perspective*, IEEE Press, pp. 137–151.

Michalewicz, Z. and Xiao, J. (1995), "Evaluation of Paths in Evolutionary Planner/Navigator", *Proceedings of the 1995 International Workshop on Biologically Inspired Evolutionary Systems*, Tokyo, Japan, pp. 45–52.

Michalewicz, Z. (1996), *"Genetic Algorithms + Data Structures = Evolution Programs"*, Springer-Verlag Berlin Heidelberg New York.

Mishra, S., Kang, M.-W., and Jha, M. K. (2014), "An Empirical Model with Environmental Considerations in Highway Alignment Optimization", *Journal of Infrastructure Systems*, 20(4), pp. 1–12.

Moavenzadeh, F., Becker, M., and Parody, T. (1973), *"Highway Cost Model Operating Instructions and Program Documentation"*, Report on Contract DOT-OS-00096, Massachusetts Institute of Technology, Cambridge, Massachusetts.

Morales, A. K. and Quezada, C. V. (1998), "A Universal Eclectic Genetic Algorithm for Constrained Optimization", *Proceedings of Sixth European Congress on Intelligent Technique & Soft Computing*, EUFIT'98, pp. 518–522.

Mortenson, M. E. (1997), *Geometric Modeling*, 2nd Edition, John Wiley & Sons, Inc., New York.

Muhlenbein, H. (1992), "Parallel Genetic Algorithms in Combinatorial Optimization", in *Computer Science and Operations Research*, edited by Balci, O., Sharda, R., and Zenios, S., Pergamon Press, New York, pp. 441–456.

Murchland, J. D. (1973), "Methods of Vertical Profile Optimisation for an Improvement to an Existing Road", *PTRC Seminar Proceedings on Cost Models and Optimisation in Highways* (Session L12), London.

Nakano, R. (1991), "Conventional Genetic Algorithm for Job Shop Problems", *Proceedings of the Fourth International Conference on Genetic Algortihms*, Morgan Kaufmann Publishers, pp 474-479.

Nicholson, A. J., Elms, D. G., and Williman, A. (1976), "A Variational Approach to Optimal Route Location", *Highway Engineers*, 23, pp. 22–25.

O'Connor, C. (1971), *Design of Bridge Superstructures*, John Wiley & Sons, Inc.

OECD (1973), "*Optimization of Road Alignment by the Use of Computers*", Organization of Economic Cooperation and Development, Paris.

Olsen, A. L. (1994), "Penalty Functions for the Knapsack Problem", *Proceedings of the First IEEE Conference on Evolutionary Computation*, IEEE Press, pp. 554–558.

ORNL (2001), "*1995 National Personal Travel Survey Databook*", ORNL/TM-2001/248, Oak Ridge National Laboratory, Oak Ridge, Tennessee.

Orvosh, D. and Davis, L. (1993), "Shall We Repair? Genetic Algorithms, Combinatorial Optimization and Feasibility Constraints", *Proceedings of the Fifth International Conference on Genetic Algorithms*, Los Altos, CA, Morgan Kauffman Publishers, pp. 650.

Orvosh, D. and Davis, L. (1994), "Using a Genetic Algorithm to Optimize Problems with Feasibility Constraints", *Proceedings of the First IEEE Conference on Evolutionary Computation*, IEEE Press, pp. 548–553.

Palmer, C. C. and Kershenbaum, A. (1994), "Representing Trees in Genetic Algorithms", *Proceedings of the First IEEE International Conference on Evolutionary Computation*, pp. 27–29.

Pareto, V. (1906), *Manual of Political Economy*, Augustus M. Kelley, New York

Parker, N. A. (1977), "Rural Highway Route Corridor Selection", *Transportation Planning and Technology*, 3, pp. 247–256.

Pearman, A. D. (1979), "The Structure of the Solution Set to Network Optimization Problems", *Transportation Research Part B*, 8, pp. 11–27.

Persaud, B. N. (1991), "*Estimating Accident Potential of Ontario Road Sections*", Transportation Research Record No. 1327, pp. 47–53.

Poch, M. and Mannering, F. (1998), "Negative Binomial Analysis of Intersection-Accident Frequencies", *Journal of Transportation Engineering*, 122 (2), pp. 105–113.

Polle, D. and Mackworth, A. (2017), *Artificial Intelligence: Foundations of Computational Agents*, 2nd Edition, Cambridge Universtity Press, 2017. Available at: http://artint.in fo/index.html.

Poole, M. R. and Cribbins, P. D. (1983), "Benefits Matrix Model for Transportation Project Evaluation", *Transportation Research*, Record No. 931, pp. 107–114.

Poorzahedy, H. and Turnquist, M. A. (1982), "Approximate Algorithms for the Discrete Network Design Problem", *Transportation Research Part B*, 16, pp. 45–55.

Pu, H., Song, T., Schonfeld, P., Li, W. Zhang, H., Hu, J., Peng, X., and Wang, J. (2019a), "Railway Alignment Optimization in Mountainous Regions through a Stepwise & Hybrid PSO Incorporating GA Operators", *Applied Soft Computing*, 78, pp. 41–57.

Pu, H., Song, T., Schonfeld, P., Li, W., Zhang, H., Wang, J., Hu, J., and Peng, X. (2019b), "A Three-Dimensional Distance Transform for Optimizing Constrained Mountain Railway Alignments", *Computer-Aided Civil and Infrastructure Engineering*, 34, pp. 972–990.

Pu, H., Zhang, H., Li, W., Xiong, J., Hu, J., and Wang, J. (2019c), "Concurrent Optimization of Mountain Railway Alignment and Station Locations using a Distance Transform Algorithm", *Computer and Industrial Engineering*, 127, pp. 1297–1314.

Pushak, Y., Hare, W. and Lucet, Y. (2016), "Multiple-path Selection for New Highway Alignments using Discrete Algorithms", *European Journal of Operational Research*, 248(2), pp. 415–427

Puy Huarte, J. (1973), "OPYGAR: Optimisation and Automatic Design of Highway Pro-files", *PTRC Seminar Proceedings on Cost Models and Optimisation in Highways* (Session L13), London.

Razieh B., Ramin, N., Ismael, G., and Zahr, M. (2018), "Forest Road Profile Optimization Using Meta-Heuristic Techniques", *Applied Soft Computing*, 64, pp. 126–137.

ReVelle, C. S., Whitlatch, E. E., and Wright, J. R. (1997), *Civil and Environmental Systems Engineering*, Prentice Hall, New Jersey.

Rich, E. (1988), *Artificial Intelligence*, McGraw-Hill, Inc., New York.

Richardson, J. T., Palmer, M. R., Liepins, G., and Hilliard, M. (1989), "Some Guide-lines for Genetic Algorithms with Penalty Functions", *Proceedings of the Third International Conference on Genetic Algorithms*, Morgan Kaufmann Publishers, pp. 191–197.

Robinson, R. (1973), "Automatic Design of the Road Vertical Alignment", *PTRC Seminar Proceedings on Cost Models and Optimisation in Highways* (Session L19), London.

Sarma, K. and Adeli, H. (2000), "Fuzzy genetic algorithm for optimization of steel struc-tures", *Journal of Structural Engineering, ASCE*, 126(5), pp. 596–604.

Sayed, T. and Rodriguez, F. (1999), "Accident Prediction Models for Urban Un-signalized Intersections in British Columbia", *Transportation Research,* Record No. 1665, pp. 93–99.

Safronetz, J. D. and Sparks, G. A. (2003), "Project-Level Highway Management Model for Secondary Highways in Saskatchewan, Canada", *Transportation Research,* Record No. 1819, pp. 297–304.

Schoenauer, M. and Michalewicz, Z. (1996), "Evolutionary Computation at the Edge of Feasibility", *Proceedings of the Fourth Conference on Parallel Problem Solving from Nature*, Springer-Verlag, pp. 245–254.

Schoenauer, M. and Michalewicz, Z. (1998), "Sphere Operators and Their Applicability for Constrained Parameter Optimization Problems", *Evolutionary Programming VII: Proceedings of the Seventh Annual Conference on Evolutionary Programming*, LNCS 1447, Springer-Verlag, pp. 241–250.

Schwefel, H.-P. (1981), *Numerical Optimization for Computer Models*, Wiley, Chichester, UK.

Shafahi, Y. and Bagherian, M. (2013), "A Customized Particle Swarm Method to Solve Highway Alignment Optimization Problem", *Computer-Aided Civil and Infrastructure Engineering*, 28, pp. 52–67.

Sharma, S. and Mathew, T. V. (2007), "Transportation Network Design with Emission Pricing as a Bi-level Optimization Problem", *Proceedings Transportation Research Board*, 87th Annual Meeting, TRB ID: 07-2143.

Shaw, J. F. B. and Howard, B. E. (1981), "Comparison of Two Integration Methods in Transportation Routing", *Transportation Research*, Record No. 806, pp. 8–13.

Shaw, J. F. B. and Howard, B. E. (1982), "Expressway Route Optimization by OCP", *Journal of Transportation Engineering*, 108(TE3), pp. 227–243.

Sheffi, Y. (1984), *Urban Transportation Networks: Equilibrium Analysis with Mathematical Programming Methods*, Prentice-Hall, Inc., New Jersey.

Siedlecki, W. and Sklansky J. (1989), "Constrained Genetic Optimization via Dynamic Reward-penalty Balancing and its Use in Pattern Recognition", *Proceedings of the Third International Conference on Genetic Algorithms*, pp. 141–150, Morgan Kaufmann Publishers Inc., San Francisco, CA, United States.

Singh, R. (1995), "Beyond the BPR Curve: Updating Speed-Flow ad Speed-Capacity Relationships in Traffic Assignment", *The Fifth Conference on Transportation Planning Methods Applications*, Seattle, Washington.

Skabardonis, A. and Dowling, R. G. (1997), "Improved Speed-Flow Relationships for Planning Applications", *Transportation Research*, Record No. 1572, pp. 18–23.

Smith, A.E. and Tate, D. (1993), "Genetic Optimization Using a Penalty Function", *Proceedings of the Fifth International Conference on Genetic Algorithms*, Morgan Kaufmann, pp. 499–503.

Smith, A. E. and Coit, D. W. (1997), "Constraint Handling Techniques — Penalty Functions", in *Handbook of Evolutionary Computation*, Chapter 5.2, edited by Back, T., Fogel, D. B., and Michalewicz, Z., Oxford University Press, pp. C5.2:1–C5.2:6.

Song, T., Pu, H., Schonfeld, P., Li, W., Zhang, H., Ren, Y., Wang, J., Hu, J. and Peng, X. (2019), "A Parallel Three-dimensional Distance Transform for Railway Alignment Optimization Using OpenMP", *Journal of Transportation Engineering, Part A: Systems*, Oct. 2019.

Song, T., Pu, H., Schonfeld, P., Li, W., Zhang, H., Ren, Y., Wang, J., Hu, J. and Peng, X. (2020), "Parallel Three-dimensional Distance Transform for Railway Alignment Optimization Using OpenMP", *Journal of Transportation Engineering, Part A: Systems*, 146(5). https://doi.org/10.1061/JTEPBS.0000344.

Steele, E. J., Lindley, R. A., and Blanden, R. V. (1998), *Lamarck's Signature. How Retrogenes Are Changing Darwin's Natural Selection Paradigm*, Perseus Books, Massachusetts.

Steenbrink, A. (1974), "Transport Network Optimization in the Dutch Integral Transportation Study", *Transportation Research Part B*, 8, pp.11–27.

Tate, M. D. and Smith, A. E. (1995), "A Genetic Approach to the Quadratic Assignment Problem", *Computers and Operations Research*, 22(1), pp. 73–78.

Teodorovic, D. (2008), "Swarm Intelligence Systems for Transportation Engineering: Principles and Applications", *Transportation Research Part C: Emerging Technologies*, 16(6), pp. 651–667.

Thangiah, S. R. (1995), "An Adaptive Clustering Method Using a Geometric Shape for Vehicle Routing Problems with Time Windows", *Proceedings of the Sixth International Conference on Genetic Algorithms*, pp. 536–543.

Thomas, R. (1991), *Traffic Assignment Techniques*, Avebury Technical.

Thomson N. R. and Sykes, J. F. (1988), "Route Selection Through a Dynamic Ice Field Using the Maximum Principle", *Transportation Research Part B*, 22(5), pp. 339–356.

Transportation Research Board (TRB) (2010), "*Highway Capacity Manual*", Transportation Research Board, Washington D.C.

Transportation Research Board (2012), "Artificial Intelligence Applications to Critical Transportation Issues", *Transportation Research Circular*, No. E-C168, Transportation Research Board of the National Academy.

Trietsch, D. (1987a), "A Family of Methods for Preliminary Highway Alignment", *Transportation Science*, 21(1), pp. 17–25.

Trietsch, D. (1987b), "Comprehensive Design of Highway Network", *Transportation Science*, 21(1), pp. 26–35.

Turner, A. K. and Miles, R. D. (1971), "A Computer-Assisted Method of Regional Route Location", *Highway Research*, Record No. 348, pp. 1–15.

U.S. DOT (1964), *"Traffic Assignment Manual"*, Bureau of Public Roads, U.S. Department of Commerce, Washington, D.C.

U.S. DOC (2000), *"National Income and Product Accounts (NIPA) of the United States for 2000"*, Bureau of Economic Analysis, U.S. Department of Commerce.

Vazquez-Mendez, M., Casal, G., and Santamarina D. (2018), "A 3D Model for Optimizing Infrastructure Costs in Road Design", *Computer-Aided Civil and Infrastructure Engineering*, 33, pp. 423–439.

Vgot, A. and Bared, J. (1998), "Accident Models for Two-Lane Rural Segments and Intersections", *Transportation Research*, Record No. 1635, pp. 18–29.

Wallace, C. E., Courage, K. G., Hadi, M. A., and Gan, A. C. (1998), *"TRANSYT-7F User's Guide"*, McTrans Center, University of Florida, Gainesville, FL, USA

Wan, F. Y. M. (1995), *Introduction to the Calculus of Variations and its Applications*, Chapman & Hall, New York.

Wang, L. Z., Miura, K. T., Nakamae, E., Yamamoto, T., and Wang, T. J. (2001), "An approximation approach of the clothoid curve defined in the interval [0, PI/2] and its offset by fee-form curves", *CAD*, 33, pp. 1049–1058.

Wang, S.-L. and Schonfeld, P. (2008), "Scheduling of Waterway Projects with Complex Interrelations", *Transportation Research Record*, 2062(1), pp. 59–65. https://doi.org/10.3141/2062-08

Webster, F. V. (1958), *"Traffic Signal Settings"*, Road Research Technical Paper No. 39, Her Majesty's Stationary Office, London.

West Virginia Division of Highways and Maryland State Highway Administration (2006), *U.S. Route 220 Tier One Environmental Impact Statement*.

Whitley, D. (2000), "Permutations", in *Evolutionary Computation 1: Basic Algorithms and Operators*", edited by Back, T., Fogel, D., Michalewicz, Z., Chapter 33.3, Institute of Physics Publishing, pp. 274–284.

Winfrey, R. (1968), *Economic Analysis for Highways*, International Textbook, Scranton, PA.

Wright, P. H. (1996), *Highway Engineering*, John Wiley & Sons, Inc., New York.

Xiao, J., Michalewicz, Z., and Zhang, L. (1996), "Evolutionary Planner/Navigator: Operator Performance and Self-Tuning", *Proceedings of the Third IEEE International Conference on Evolutionary Computation*, IEEE Press, pp. 366–371.

Xiao, J., Michalewicz, Z., and Trojanowski, K. (1997), "Adaptive Evolutionary Planner/Navigator for Mobile Robots", *IEEE Transactions on Evolutionary Computation*, 1(1), pp. 18–28.

Xiong, Y. and Schneider, J. B. (1992), "Transportation Network Design Using a Cumulative Genetic Algorithm and Neural Network", *Transportation Research*, Record No. 1364, pp. 37–44.

Yang, H. and Yagar, S. (1994), "Traffic Assignment and Traffic Control in General Freeway-Arterial Corridor Systems", *Transportation Research Part B*, 28, pp. 463–486.

Yang, H. and Lam, W. H. K. (1996), "Optimal Road Tolls under Conditions of Queuing & Congestion", *Transportation Research Part A*, 30(5), pp. 319–332.

Yang, H. and Bell, M. G. H. (1998), "Models and Algorithms for Road Network Design: a Review and Some New Developments", *Transport Reviews*, 18, pp. 257–278.

Yang, N., Kang, M.-W., Schonfeld, P., and Jha, M. K. (2014), "Multi-objective Highway Alignment Optimization Incorporating Preference Information", *Transportation Research Part C*, 40, pp. 36–48.

Yeniay, O. (2005), "Penalty Function Methods for Constrained Optimization with Genetic Algorithms", *Mathematical and Computational Applications*, 10(1), pp. 45–56.

Yin, Y. (2000), "Genetic Algorithms Based Approach for Bi-level Programming Models", *Journal of Transportation Engineering*, 126(2), pp. 115–120.

Yokota, T., Gen, M., Ida, K., and Taguchi, T. (1996), "Optimal Design of System Reliability by an Improved Genetic Algorithm", *Electronics and Communications in Japan Part III-Fundamental Electronic Science*, 79, 41-51.

Zegeer, C. V., Stewart, R., Council, F. M., Reinfurt, D. W., and Hamilton, E. (1992), "Safety Effects of Geometric Improvements on Horizontal Curves", *Transportation Research*, Record No. 1356, pp. 11–19.

Zhou, T. (2014), *"Optimization Models for Runway Location, Orientation and Longitudinal Grade Design"*, M.S. Thesis, University of Maryland.

Zhou, T. and Schonfeld, P. (2015), "A Model for Optimizing Runway Location and Orientation," *Proceedings of Transportation Research Board, 95th Annual Meeting*, TRB ID: 15-3439, Washington, DC.

Zun W. (2011), *"Geometric and Environmental Considerations in Highway Alignment Optimization,"* M.S. Thesis, University of Maryland.

Index

AASHTO design manual, 37
AI method, 17
    summary, 29
AI methods for optimizing alignments, 17
    distance transform, 27
    genetic algorithm, 17
    heuristics, 18
    neighborhood search heuristic, 26
    PSO algorithm, 25
    simulated annealing, 17
    Swam intelligence, 17
    Swarm intelligence, 24
American Association of State Highway
    and Transportation Officials
    (AASHTO), 14, 15
artificial intelligence, 8, 17, 75

bi-level DNDP
    cost minimization, 127
    lower-level DNDP, 127
    NP-complete, 126
    traffic assignment, 127
    upper-level DNDP, 127
bi-level HAO, 8
bi-level HAO model, 131, 134, 149, 160
    applicability, 170
    concept, 132
    example road network, 152
    follower, 131
    future work, 161
    GA, 148
    GIS, 148
    GIS inputs, 149
    highway agency cost, 135

inputs, 148
intersection delay estimation, 159
leaders, 131
lower level, 133, 145
model structure, 148
modular structure, 136
network representation, 151
O/D trip matrix, 150
penalty and environmental costs,
    144
preprocessed traffic assignments,
    147
static (and deterministic) traffic
    assignment, 146
summary, 160
traffic assignment, 9, 133, 139, 150
traffic reassignment, 147
upper-level, 132
user cost saving, 135
bi-level HAO model application, 163, 211,
  220
    cost breakdown, 229
    example description, 163
    ICC Project, 213
    optimized alignments, 225
    optimized alternatives, 166
    preprocessed traffic assignments,
      221
    sensitivity analysis, 230
    summary, 169
    traffic reassignment, 224
bi-level programming problem, 125
BPR function, 157

Brookeville Bypass Study, 102, 109, 117, 174
    computation breakdown, 110
    computation time, 103
    control areas, 178
    cross-section, 179
    elevation range, 176
    environmental impact, 183
    geographically sensitive regions, 176
    land-use, 175
    optimized alignments, 188
    optimized solution, 104
    project objectives, 173
    study area, 118, 174
    total costs, 183

conventional highway design process, 3, 4
    economical path, 7
    manual cost-benefit evaluation, 3

design constraints
    horizontal sight offset, 15
    maximum allowable gradient, 71
    maximum gradient constraint, 58
    maximum gradients, 15
    minimum horizontal curvature
        constraint, 57
    minimum horizontal curve radius,
        15, 71
    minimum horizontal sight distance,
        71
    minimum horizontal sight offset, 58
    minimum length of crest and sag
        vertical curves, 16
    minimum length of spiral transition
        curve, 71
    minimum length of vertical curve, 71
    minimum length of vertical sight
        distance, 71
    minimum superelevation runoff
        length, 15, 58
    minimum vertical clearance, 16, 58
    minimum vertical curvature
        constraint, 58
digital elevation model, 43, 176, 218
digital terrain model, 43

direct constraint handling
    decoding method, 79
    elimination method, 78
    feasible regions, 80
    preserving method, 79
    repairing method, 78
discrete network design problem, 127, 128
    bi-level programming, 128
    macroscopic, 128
    objective, 128
distance transform, 27, 234
DNDP
    bi-level DNDP, 126

environmental and geographical
    constraints
        geographically sensitive regions,
            59
environmental and socio-economic
    impacts, 14
        control areas, 16
        environmentally sensitive areas, 16
        fixed points, 16
        penalty, 14
        socio-economically sensitive areas,
            16
evolutionary algorithms, 75
    constraint handling, 75
    direct constraint handling, 75, 78
    GA, 75
    indirect constraint handling, 75, 80
    penalty method, 77

feasible gates, 87, 89, 96, 175
    horizontal feasible gate, 90, 92, 94,
        99
    input data preparation module, 90
    maximum gradient, 99
    motivation, 87
    penalty, 102
    untouchable areas, 97
    vertical cutting line, 93
    vertical feasible gates, 99
FG method, 186
    example study, 102
    HFG method, 90, 92, 95

Summary, 106
VFG method, 99

GA operators, 47
    one-point crossover, 47
    two-point crossover, 47
    uniform mutation, 47
GA-based HAO model, 24, 25, 31, 71, 84,
    109, 121, 160
    chromosomes, 33
    computation efficiency, 87, 109
    constraint handling, 84
    endpoint determination, 45
    endpoint generation, 47
    extension, 123
    FG method, 87, 102
    fitness, 33
    generation, 33, 39
    genetic operators, 34, 35
    GIS computation, 111
    GIS module, 110
    highway endpoints, 42
    indirect constraint handling, 84
    infeasible solutions, 84
    jointly, 40
    objective function value, 34
    offspring, 33, 47
    optimization process, 34
    orthogonal cutting planes, 34
    other models, 23
    penalty method, 84
    PI density, 36
    pool-based search, 33
    P&R method, 109, 112
    real-world road project, 102
    representation of highway
      alignments, 34
    reproduce, 33
    the number of PI's, 36
    UMD Research Team, 18
genetic algorithm, 8, 17, 31, 75, 78, 81
    constraint handling methods, 18
    evolution, 17
    fitness function, 17
    genetic operators, 18
    key advantages, 18

penalty, 81
pool-based search, 18
survival of the fittest, 17
geographic information system, 8
goodness test, 185, 231

HAO 1.0
    cost function, 21
    three-dimensional, 18
HAO 2.0
    GIS analysis, 21
HAO 3.0
    structure costs, 22
HAO 4.0
    Feasible Gate, 22
    Prescreening & Repairing, 22
HAO 5.0
    sight distance module, 22
    vehicle dynamic module, 22
HAO 6.0
    Hybrid Multi-Objective Genetic
      Algorithm, 22
HAO 7.0
    bi-level, 22
    road network, 22
HAO model, 8, 9, 18, 23, 90, 109, 235
    Bi-level HAO, 134
    GA and GIS communicate, 59
    GA-based HAO model, 8, 9
    GA-based optimization, 60
    GA-GIS based HAO model
      structure, 61
    GIS module, 90
    GIS-based evaluation, 60
    GIS-based model, 29
    HAO 1.0, 18, 42
    HAO 2.0, 21
    HAO 3.0, 22
    HAO 4.0, 22
    HAO 5.0, 22
    HAO 6.0, 22, 42
    HAO 7.0, 22, 42
    inputs, 60
    jointly optimize, 18
    modular structure, 70
    penalty, 71

potential improvements, 235
real-world applications, 173
UMD research team, 28
HAO model application, 173, 195
    Brookeville Bypass project, 173
    computation efficiency, 183
    goodness of solutions, 186
    inputs, 178
    map digitization, 174
    optimized alignments, 206
    optimized horizontal alignments, 181
    optimized vertical alignments, 182
    outputs, 179
    sensitivity analysis, 188
    US 220 project, 200
HAO model constraints, 57
    constraints, 85
    design constraint violation, 115
    design constraints, 57, 71
    environmental and geographical
        constraints, 57, 59
    environmental cost, 73
    life-cycle cost, 74
    penalty, 72
HAO model cost functions, 55, 57, 62
    accident cost, 69
    earthwork cost, 64
    length-dependent cost, 62
    maintenance cost, 66
    penalty cost, 70
    right-of-way cost, 63
    structure cost, 65
    travel time cost, 68
    user cost, 68
    vehicle operating cost, 69
HAO model extensions, 235
    railway alignments, 234
HAO model formulation, 55
HAO model objective function, 55
HAO model-related developments, 233
highway alignment, 6, 33
    horizontal alignment, 33
    transition curve, 15
    vertical alignment, 33
highway alignment optimization, 4, 7, 9,
    127, 129

computational burden, 130
constraints, 8
factors, 4
highway costs, 8
multi-objective optimization, 28
optimization methods, 6
trade-off analysis, 14
traffic assignment, 130
highway capacity manual, 157
highway constraints, 15
    design constraints, 15, 16
    geographical and environmental
        constraints, 16
highway cost, 11
    alignment-sensitive costs, 11
    construction costs, 12
    dominating costs, 11
    earthwork cost, 11, 12
    environmental and socio-economic
        impacts, 14
    highway agency cost, 12
    length-dependent cost, 12, 13
    location-dependent cost, 12
    maintenance costs, 13
    right-of-way cost, 11, 12
    user costs, 13
    volume-dependent cost, 12
highway cost functions
    highway agency cost, 62
highway structures, 49, 51, 65
    bridges, 49
    crossing rivers, 49
    grade separation, 49, 54
    interchange, 54
    intersections, 52
    retaining walls, 50
    simple trumpet interchange, 52
    small highway bridges, 50
    structure cost, 51
    three-leg at-grade intersection, 52
    tunnels, 49
horizontal alignment, 37
    circular curve, 37
    horizontal curve, 34, 36
    tangents, 37

indirect constraint handling
  penalty method, 81
input data preparation module, 98
intersection delay estimation
  HCM models, 159

life-cycle cost, 6, 13, 74

maintenance costs, 13
Maryland ICC Project, 211
  alignment search space, 217, 221
  assumptions and limitations, 213
  existing road network, 214
  geographic entities, 216
  ICC impact area, 216
  input parameters, 218
  objective function, 213
  optimization results, 220
  optimized alignment, 226
  preprocessed traffic assignments,
    221
  project description, 211
  purposes, 212
  study area, 212
  traffic analysis zones, 214
  traffic information, 214
  traffic reassignment, 224
Maryland Intercounty Connector, 211
Maryland State Highway Administration,
  109, 173, 195
MdProperty View, 102, 174, 197, 217
modeling highway endpoints, 42
  3D highway endpoint, 46
  assumption, 42
  at-grade intersection, 46
  endpoint determination, 44
  grade separated structure, 46
  highway endpoints, 42
  reference points, 46
modeling highway structures, 49
  at-grade intersections, 54
  grade separation, 50
  reference points, 54
  small highway bridges, 50
modeling horizontal alignments, 36
  circular curves, 36

horizontal alignment generation
  procedure, 39
  minimum radius, 37
  minimum superelevation run-off
    length, 37
  spiral transition curve, 37
  transition curves, 36
modeling vertical alignments, 40
  design constraints, 41
  ground profile, 41
  piecewise linear trajectory, 41
  Road Elevation Determination
    Procedure, 41
multi-objective optimization
  Pareto Front, 28
  tradeoffs, 28

National Environmental Policy Act, 16
neighborhood search-heuristic, 26
network design problem, 125, 126, 129
  CNDP, 125
  DNDP, 125

orthogonal cutting planes, 39, 89, 99

P&R method
  advantageous, 120
  computation time savings, 119
  curve fitting violation, 113
  example study, 117
  horizontal alignment evaluation,
    113
  key idea, 121
  penalty cost, 117
  prescreening and repairing, 186
  summary, 121
Particle Swarm Optimization, 24
penalty method, 80
  adaptive penalty, 84
  computational burdens, 84
  death penalty, 81
  dynamic penalty, 83
  soft penalty, 82
  static penalty, 82
points of intersections, 21, 33, 37, 89
  horizontal PI, 34

HPI, 34
PI density, 36
vertical PI, 35
VPI, 35
prescreening and repairing, 109, 111, 112
        concept, 112
        motivation, 109

soft penalty function, 71, 117

traffic assignment, 145
        link-performance functions, 156
        system optimal, 145
        user optimal, 145
travel time cost saving
        steps, 140

uniform mutation, 47
US 220 Project, 195
        design criteria, 196
        elevation map, 199
        Geographical Constraints Map, 200
        geometric design specification, 203
        GIS data, 197
        ground elevation, 200
        HAO model application, 196
        land cost map, 198

land-use map, 197
new bypasses, 203
objective function, 207
optimize alignments, 205
priority funding areas, 195
project description, 195
study area, 196
total cost breakdown, 209
typical section, 207
Untouchable region, 202
user cost saving
        concept, 137
        travel time cost saving, 138
        vehicle operating cost saving, 141
user costs, 13
        accident cost, 13
        travel time cost, 13
        vehicle operating costs, 13

vehicle operating cost saving
        steps, 143
        vehicle depreciation cost, 141
vertical alignment, 40, 99
        fitting parabolic curves, 40
VFG method
        representation, 93, 100

www.ingramcontent.com/pod-product-compliance
Lightning Source LLC
Chambersburg PA
CBHW050546190326
41458CB00007B/1941